MAY	JUNE	JULY	AUG	SEPT	OCT	NOV	DEC

How to Grow
Plants
from
Seeds

RHS How to Grow Plants from Seeds

Authors: Sophie Collins & Melissa Mabbitt

First published in Great Britain in 2021 by Mitchell Beazley, a division of
Octopus Publishing Group Ltd
Carmelite House, 50 Victoria Embankment, London EC4Y 0DZ
www.octopusbooks.co.uk

An Hachette UK Company

www.hachette.co.uk

Published in association with the Royal Horticultural Society

ISBN: 978-1-78472-762-8

A CIP record of this book is available from the British Library

Printed and bound in China

Conceived, designed and produced by The Bright Press
an imprint of The Quarto Group
The Old Brewery, 6 Blundell Street,
London N7 9BH, United Kingdom
T (0) 20 7700 6700
www.QuartoKnows.com

Publisher: James Evans

Art Director: James Lawrence

Editorial Director: Isheeta Mustafi

Managing Editor: Jacqui Sayers

Project Editor: Polly Goodman

Design: Wayne Blades

Illustrations: Melvyn Evans

Mitchell Beazley Publisher: Alison Starling

RHS Publisher: Rae Spencer-Jones

RHS Consultant Editor: Simon Maughan

RHS Head of Editorial: Tom Howard

The Roy   ity
dedicated to adva   ritable work
includes providing ex   of gardeners,
creating hands-on opp   earch into plants,

For mo   6.

How to Grow
Plants from Seeds

· ·

Sowing seeds for flowers,
vegetables, herbs and more

SOPHIE COLLINS
& MELISSA MABBITT

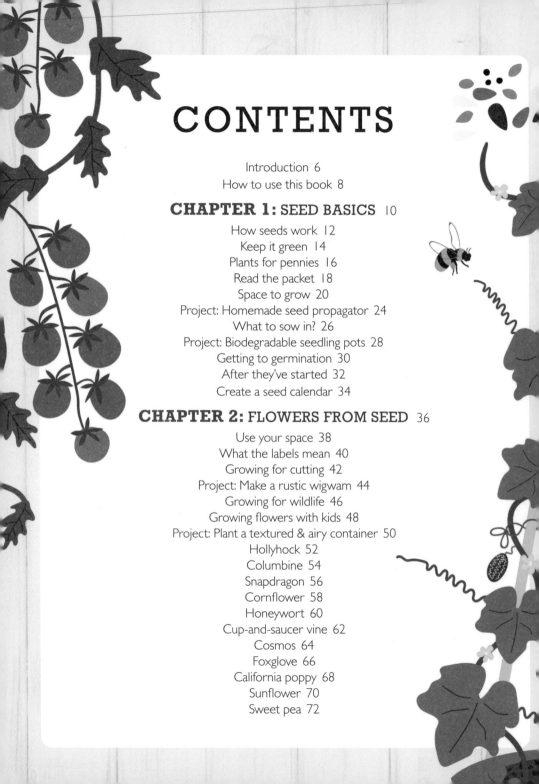

CONTENTS

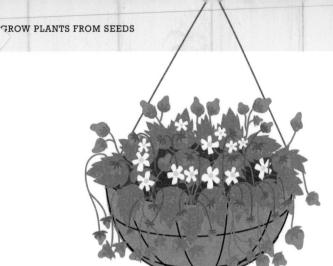

INTRODUCTION

Watching plants grow from seeds you've sown yourself is the best sort of everyday magic. You're undertaking a true collaboration with nature. With a bit of nurturing, you can wake seeds from their dormancy and bring new plants to life, and if you plan ahead, you can put together combinations of plants to suit your own style and circumstances. These may be containers planted up with gloriously colourful flowers, a productive vegetable garden, or the transformation of a dull yard into something vibrant and lovely.

If you've previously depended on young plants bought from professional nurseries, be assured that growing plants from seeds is the simplest skill to learn. It's a short step from germination to strong little plants, which, just a few weeks down the line, will not only give you flowers and fruit, but all the satisfaction that successfully growing your own brings, too.

How to Grow Plants from Seeds will hold your hand through the process, guiding you all the way from which seeds to buy and how to sow them, to how to nurture your seedlings through to productivity. It includes projects to try, and in-depth plant profiles for the best flowers and vegetables to start off with. Finally, rounding off, it shows you how to collect the seeds of the plants you've grown, so that you can keep the cycle of your new enthusiasm going.

One other thing that starting plants from seeds offers – it's full of surprises. As with any new enterprise, you won't be 100 per cent successful with everything you try, but there's no pleasure like that of raising productive plants from seeds that a friend told you were 'tricky', or nursing a batch of seedlings that accidentally caught a cold. Seed-growers are a club of the best people; this book is your membership card.

HOW TO USE THIS BOOK

How to Grow Plants from Seeds delivers on the promise – even if you're a complete novice, it offers all the information you need to start to grow your own plants successfully from seeds.

CHAPTER 1

Covers the fundamentals of how seeds work, with detailed descriptions of how to sow them, and how to raise healthy seedlings that are right for your growing space, however large or small it is. Includes projects on how to craft simple seedling pots from paper, and how to make your own propagator.

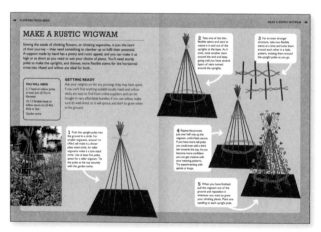

CHAPTER 2

Explains how to grow flowers from seeds, from scratch. Looks at how to choose from the huge selection available, what the labels on the packets mean, how to use your space to the best advantage, and how to start your own cutting garden. Includes projects on how to make a rustic wigwam for climbers, and growing an inspirational container.

CHAPTER 3

A primer in growing vegetables from seeds, covering everything from timing your sowing, to getting the best crops, and growing your own Halloween pumpkin. Also included is a project on growing a container of edibles that looks as good as the vegetables will taste.

CHAPTER 4

Takes your interest as a gardener on to the next step, showing how you can gather seeds from the plants you've grown and repeat the cycle next year. Plus a look at seedswapping events and how they work.

PLANT PROFILES Included in chapters 2 and 3 are individual profiles of 15 flowers and 15 vegetables, including three cultivars or similar varieties to try, offering in-depth information on some of the best choices for growing your own when you're starting out.

Finally, the book ends with a glossary, explaining technical terms, and a resources section including seed retailers, websites, books and other inspirational sources that will help you get going with growing plants from seeds.

CHAPTER ONE

SEED
BASICS

If you're fairly new to gardening, you may find the idea of growing from seeds a bit more daunting than simply buying seedlings or plants. Starting seeds off, though, is far easier than you might think. They already contain everything they need to grow – they're hard-wired to produce new plants. All you have to do is give them a little help. And once seeds have grown into seedlings, it's equally straightforward to supply the few things – water, the right soil, enough light – that will see them through to healthy adulthood.

Seed Basics gives you the background you need to get off to a successful start. It looks at how germination works, and at why seeds are the best option both in economical and environmental terms. It then takes you through the processes step-by-step – from sowing, to seedling, to plant. It covers the questions new gardeners ask, and reminds you to keep a record so that you can plant your gardening year effectively, whether you're planting up a window box, a terrace full of containers, or a full-sized garden.

HOW SEEDS WORK

The seeds you buy are dormant – they have the potential to grow, but they're not active. Every healthy seed, though, holds all the ingredients it needs to germinate, and if it's offered the necessary basics of oxygen, moisture and warmth, it will sprout. Requirements vary a little – some seeds need light to germinate and some don't, for example – but the vast majority conform to some basic rules.

HOW SEEDS START

Planting seeds at the right time, watering them, and keeping them at a fairly even temperature, is usually enough to get things going. Every seed contains its own energy pack, the cotyledon, which will feed it as it germinates. As moisture starts to soften the seed's outer shell, the coating breaks open and the two halves of the cotyledon separate. The plant first puts out a rootlet, which grows down into the soil, followed shortly afterwards by a small shoot growing upwards. The shoot breaks through the soil, straightens, and soon produces its first leaves at the tip. In some plants,

the cotyledon is carried above the soil by the shoot – you'll sometimes see it as two rounded halves on either side of the shoot – but it's not to be confused with the first proper leaves, also called the 'true' leaves, which follow on later. In others, the cotyledon stays below ground, but carries on feeding the seedling.

Straight after germination, the seed relies on its own food stores until it produces its first leaves.

GROWING LARGER

Once its first leaves have unfurled and grown, the plant is no longer dependent on its food stores because it can begin to photosynthesize, the process by which it will feed itself. Photosynthesis takes place in the leaves; it's a chemical reaction that allows the plant to convert light and moisture into energy. Once the seed is self-sufficient, it has become a seedling.

Despite the fact that the seeds of various plants may look very different from one another – think of the minute specks of foxgloves, the solid little spheres of sweet peas and the kernels of sweetcorn, for example – and they may have varying preferences when it comes to soil, or temperature or light levels, they all follow roughly the same path to germination.

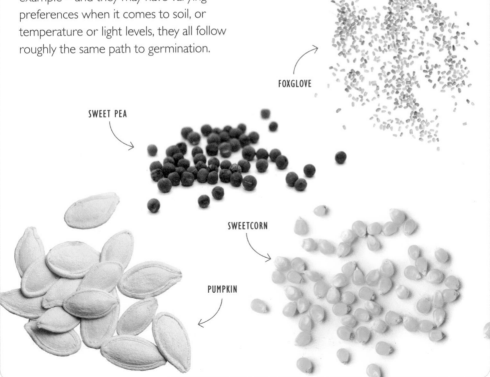

FOXGLOVE

SWEET PEA

SWEETCORN

PUMPKIN

KEEP IT GREEN

At a time when concern for the environment is at an all-time high, growing from seed makes good green sense. You'll have the widest choice of options when it comes to choosing what to buy, and control of your growing methods and medium. Best of all, you can be sure they have been produced without chemicals: great both for you, especially if you're growing things to eat, and for the wildlife in your garden.

BIGGER ISN'T BETTER

Buying larger plants is rarely the green choice. The vast majority, from tiny plugs to full-sized adults plants, whether sold online or in nurseries, supermarkets and other outlets, are raised intensively. Many are grown using huge amounts of plastic, and raised in composts containing peat, an environmental bugbear because peat is taken from some of the most endangered habitats anywhere. These plants are closely guarded to ensure that they look green and perky enough to please buyers, so they've

also often been subjected to regimes of intensive spraying and feeding before you bring them home.

You might, for example, buy a beautiful foxglove, in bud and ready to flower, that has already been sprayed with insecticide, but that is nevertheless labelled 'bee-friendly'. In recent studies which tested plant samples bought from commercial nurseries in the UK and Europe, many were found to contain an enduring chemical residue which, when they were relocated to gardens, remained at levels high enough to harm pollinators and other insects.

The serried ranks of healthy looking plants you see at the nursery may have been grown without much attention to green issues.

FOXGLOVE

BEST OF BOTH WORLDS

Seeds produced for home gardeners don't carry this load. They won't harm wildlife and if, like many gardeners, you'd prefer to manage your plot without using heavy-duty chemicals, you'll be starting as you mean to go on. You can choose a green growing medium to start seeds off, too, avoiding peat-based or fertilizer-'enriched' composts.

When it comes to choosing what to grow, the range of seeds available is much broader than you'll find on offer as young plants in a nursery, so you can be very specific in choosing not only your favourites, but also those that will fit your growing space best and do what you want them to. This applies whatever you're looking for – plants that will work in full sun, or shade, or seeking out the most pollinator-friendly choices, or cultivars that are particularly suited to pots.

PLANTS FOR PENNIES

Not only is growing from seed greener, it's also cheaper. The price of seeds varies according to how rare or difficult to harvest they are, and the number of seeds in each packet varies a lot, but even the most expensive seeds include more than a few to a packet, while there can be hundreds of the smaller ones, all for an outlay of just a few pounds.

HOW MUCH IS ENOUGH?

The most common mistake you can make when you're new to seeds is overbuying. Unless you're growing on a huge scale, a single packet of each variety is plenty for your year's needs, and though you can save spare seeds to plant next year, you'll usually get the best germination rates from fresh seeds. On the other hand, seeds are cheap enough for you to experiment, so if you want to choose more variety than you need, go ahead – if you end up growing more plants than you can house, fellow gardeners will be happy to take the surplus off your hands.

RUNNER BEANS

CONVENIENCE SEEDS

Pelleted seeds and seed tape have been designed to make seed-sowing easier. Pelleted seeds have a pill-like coating that makes tiny seeds easier to handle. They must be watered in thoroughly to ensure the coating dissolves. Seed tapes have seeds encased in them at regular intervals. You measure the tape into a furrow, then cover it with soil, leaving the seeds spaced and sown, and the tape to gradually dissolve. Both products are convenient, although the convenience comes at a slightly premium price.

SEED TAPES

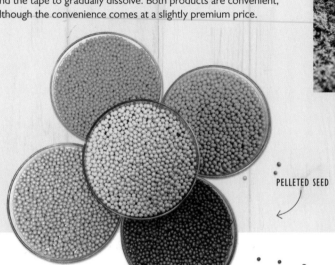

PELLETED SEED

WHERE TO BUY

Seeds are available in so many places – supermarkets and nurseries, or online, produced by both large companies and small independents – that it can be hard to know where to start. The plus side of looking at real packets is the hands-on joy of browsing. Buying online, though, has its own rewards: while you can't actually handle the packets, you'll find all the same facts, sometimes organized in a more accessible way, with groups of plant types offered with introductions and supplementary information. The smaller, independent producers may have an online-only presence and smaller ranges to offer, but they're worth looking at; they often have the most characterful descriptions of their seeds, and sometimes stock rarer or more interesting varieties.

Finally, there are catalogues; a number of seed firms still offer them, and once you're on the first couple of mailing lists you'll find that others start dropping through the door. They make good winter-evening reading as you plan your seed-growing for next year.

READ THE PACKET

Seed packets deserve more than a casual look. They're full of information, although it's presented small-scale and in various different ways, so you may have to squint a little to figure them out. Here are key things to watch out for, and what they are telling you.

The information given on seed packets varies, depending on which manufacturer they're from, but most packets will show at least the following:

1. NAME

The front of the packet often shows both the common and the Latin names of the plant, with a photograph. It will tell you, with the notation 'F1', if the plant is a hybrid (see page 41).

It will often also identify the plant as an annual, biennial or perennial (see page 40).

2. NUMBER OF SEEDS

This tells you roughly how many seeds you'll find in the packet. It can vary from just a few of the larger seeds, especially when they're an unusual variety, to huge numbers of smaller ones – a packet of tiny seeds may say 'Average 500', for example.

3. TIMINGS

This is often shown as a year's calendar laid out in a line with the months marked. Coloured lines are keyed to tell you in which months seeds should be sown, and when they will bloom (flowers) or crop (food). It gives you an idea of how this plant will fit in with your overall growing plans.

CHARD

'BRIGHT LIGHTS'

Beta vulgaris cicla var. *flavescens*

4. SOWING INSTRUCTIONS

The basics – whether you should sow the seeds outside or under cover, and when they should go in, plus the depth they should be sown at, and how long it's likely to take them to germinate.

5. HEIGHT AND SPREAD

How tall the plant will grow and how much elbow room it will need in a bed or container.

Compare and contrast it with other things you're growing, and you'll quickly learn to think to scale when it comes to plants – so you won't plan to grow outsize plants in a tiny space or, just as annoying, find that you've grown some small and underwhelming plants to fill a large area.

6. SPACING

How much space you should leave between plants if you're planning to grow them in quantity. This one is particularly important for vegetable growing, or if you're planning a cutting garden.

7. SOW BY

This is your seeds' 'best before' date. It may be this year, or a year or two in the future, but make sure it's not already past – fresh seed has the highest germination rates.

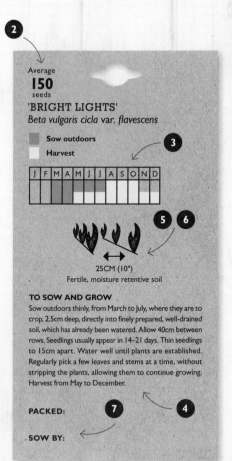

Average
150
seeds

'BRIGHT LIGHTS'
Beta vulgaris cicla var. *flavescens*

■ Sow outdoors
□ Harvest

J	F	M	A	M	J	J	A	S	O	N	D

25CM (10")
Fertile, moisture retentive soil

TO SOW AND GROW
Sow outdoors thinly, from March to July, where they are to crop, 2.5cm deep, directly into finely prepared, well-drained soil, which has already been watered. Allow 40cm between rows. Seedlings usually appear in 14–21 days. Thin seedlings to 15cm apart. Water well until plants are established. Regularly pick a few leaves and stems at a time, without stripping the plants, allowing them to continue growing. Harvest from May to December.

PACKED:

SOW BY:

CHARD SEEDS

SPACE TO GROW

When we shop for seeds, we imagine the plants grown-up, flowering in the container or bed they'll eventually grow into. But as well as planning for where they'll go later, it's also important to sort out space for the interim stages, when pots and trays of sown seeds and subsequently seedlings need somewhere to sit, with enough light and warmth until they've started to grow.

Turn seed trays around daily to stop the seedlings leaning towards the light as they grow.

INSIDE OR OUTSIDE?

Some tender seeds need to be started off 'under cover' – which can variously mean indoors, in a propagator or on a windowsill, or in a cold frame or greenhouse. Others, though fewer, need to be put directly into the ground where they're to grow, because the seedlings don't like being moved around after germination. Many varieties can be started either way – if you want them to get an early start while temperatures are still low outside you can sow them indoors; if you're happy for them to germinate and get onto the seedling stage a bit later, you can sow them outside a month or two later, as the weather starts to warm up and the danger of a late frost fades.

KNOW WHAT YOU'VE GOT

When you've collected all the seeds you want for the season, look at them together, reading the packets and noting what you've got. How many seeds need to be started outside and when? How many need to be under cover? And how warm do they like to be? Make a list, so you know how many pots and trays you'll be housing and for how long. Assume you'll be sowing around 20 per cent more seeds than the number of plants you'd like, to allow for occasional failures in germination and natural wastage in seedlings. Although you don't have to plan it down to the last seedling, it's helpful to have a rough picture of what space you'll need and when.

Propagators are available in all sizes – the small ones are useful if you want to protect one or two pots of special seeds.

INDOOR PROTECTION

Pots and trays can be put on windowsills or, if they won't be disturbed, on the floor in a corner that gets plenty of light. Seeds and just-germinated seedlings prefer fairly steady temperatures, so try to avoid leaving them anywhere that will get too hot (a windowsill in full sun that's also near a heated radiator, for example) or, equally, where they may be affected by cold draughts. A propagator can be useful to maintain moisture and warmth. At its simplest, this is just a box with a clear, vented lid, although more complicated and pricier versions with inbuilt watering and heating systems are available. You can also make your own (see pages 24–25). And it's possible to improvise by loosely wrapping a plant pot or tray in a clear plastic bag if you have some seeds that need more warmth and humidity than they are getting in order to germinate.

A cold frame or greenhouse – whatever its size – is useful year round, not just in sowing season.

TEMPORARY MEASURES

If you have more space for pots and trays outside than inside, a cold frame – a raised frame with a hinged glass or plastic lid, but no base – will house seeds that don't need much warmth to germinate, but do need protecting from the extremes of cold or wet weather. You can also buy portable mini-greenhouses, with shelves and clear plastic or glass walls. They come in lots of different sizes and at a range of prices. Some are small enough to fit on the smallest patio, terrace or even balcony, and they're useful for every stage of home growing, from housing seeds and young plants, to offering shelter to adult plants such as tomatoes, which aren't hardy enough to manage spells of unseasonably cold weather. Either a cold frame or a portable greenhouse is a good investment for any gardener; you'll get years of use out of them, and both have lots of different functions. And of course, if you're lucky enough to have a larger greenhouse, you'll be using it to capacity at sowing season.

THE GREAT OUTDOORS

Seeds that are sown outdoors go into the ground later so, in theory at least, the weather should be warmer, but they may still appreciate some extra protection from cloches. Again, these come in many forms, from beautiful (and expensive) Victorian designs in iron and glass, to simple bell cloches

in either glass or plastic. At its most basic, though, an effective cloche can be made from a cut-down, clear plastic bottle with the lid left off – push the cut rim into the ground around the point where the seed or seeds have been sown. After the seed germinates, you can take the cloche off during the day, but replace it at night when it will do double duty, protecting from both plummeting temperatures and voracious slugs.

HOMEMADE SEED PROPAGATOR

While some seeds need low temperatures to prompt germination, others need steady warmth to start to grow. Propagators that keep seeds at a steady temperature are really useful, but they can be expensive. Here's an easy way to make a homemade version.

YOU WILL NEED

2 clear, plastic fruit punnets

Packing tape

Compost

Toilet-roll tubes (optional)

Egg box (optional)

Tray

GETTING READY

Clear, plastic fruit punnets, such as the ones you buy grapes or tomatoes in at the supermarket, are ideal for this project. You can use them to make a miniature greenhouse for your windowsill, which will create a cosy environment for seeds to germinate and protect tender seedlings from drying out. If you have a small thermometer, you could pop this inside, too, to make regulating the temperature inside even easier.

1 Place one punnet upside down on top of the other, edge to edge, and run some packing tape along one edge so the punnets open and shut like a clam shell. This will make your basic, protective environment, but there are a few options as to how you fill it.

2 The simplest way to use the propagator is to half-fill it with compost and sow seeds directly inside; water, then flip on the 'lid' and wait for the seeds to germinate. Prick out the seedlings into pots once they are large enough to handle.

3 Alternatively, you can use toilet-roll tubes or egg boxes to make modular inserts. As they are made of soft cardboard, once the seedlings have grown, they can be lifted out and planted outside, where the cardboard will disintegrate into the soil. To make the modules, place six or seven toilet-roll tubes or an egg box into the bottom punnet, making sure they are a snug fit. Fill each module with compost and firm down with your fingers; then sow the seeds.

4 Place three or four of these mini-propagators on a tray to catch the drips when watering, and place the tray on a windowsill. On sunny days, when temperatures rise it can quickly get too hot for most seedlings, so flip off the lids to stop the heat building up, and replace them as evening arrives.

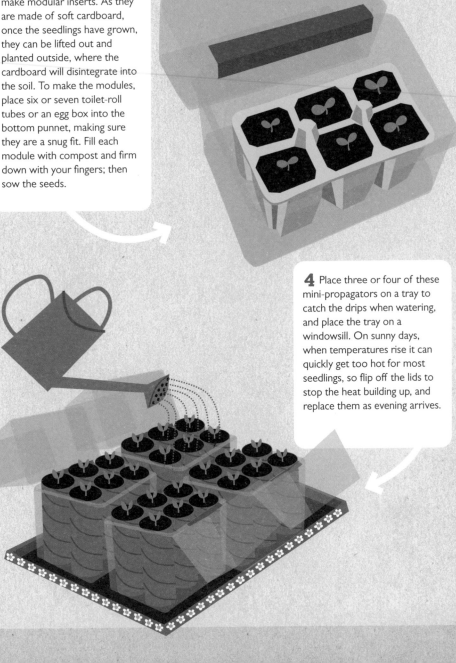

WHAT TO SOW IN?

When you've chosen your seeds and checked the packets for any dos and don'ts, it's time to start sowing. There are a lot of different containers for raising seedlings, from the traditional, small, plastic pots to greener, more environmentally friendly options – and of course there are also some that you can improvise or recycle from materials you've already got.

STARTING OFF

Start seeds off in small pots, around 9cm (3.5in) diameter, or seed trays divided into individual units called cells. Undivided seed trays are widely available, but they are a bit more challenging when it comes to potting your seedlings on – the tiny plants tend to clump together and are harder to handle when they're not already separated into smaller divisions.

WOOD-FIBRE POTS

You probably already have some plastic pots, cells and trays, because they're the go-to containers for most nurseries selling small plants or plug plants. Keep them and use them, but if you want to avoid plastic generally, there are lots of alternatives when you need to get more.

For biodegradable containers, hunt out pots and cells made from wood fibre (check the label before you buy, because some also contain peat). They're single-use – as they get wet, they gradually disintegrate – so as a seedling grows, you can put the whole pot into a larger container; the roots will simply grow through the original. Small clay pots are good, too, although you need to be careful to keep the soil in them moist because they dry out easily. Coir pellets are another option; they arrive as desiccated little discs, but soak them in water for an hour and each one plumps up into a cylinder. You pop the seeds into its top. You can make your own starter pots from newspaper, too (see pages 28–29). They're simple, quick and economical.

REUSE AND RECYCLE

If a container is about the right size and has some drainage holes in the base, it will work for seeds. This means that you can press all sorts of used items into service, from takeaway coffee containers to yoghurt pots. Be as creative as you like – the seeds will still grow.

Last but not least, big plastic or tin trays are useful for moving lots of small pots and cells around. Trays with a lip or edge make it easier to water your seeds neatly – if you don't already have them, charity shops are always a good source.

RECYCLED POT

Unlike other containers, wood-fibre pots and cells don't have drainage holes, because they are naturally porous.

BIODEGRADABLE SEEDLING POTS

Some seedlings, such as sweet peas and beans, are best grown in pots to protect them from the cold or pests, but also dislike having their roots disturbed. This poses a problem when it comes to transplanting them into the ground. These biodegradable pots will protect the roots as you transplant the seedlings, and then quickly break down and disappear into the soil.

YOU WILL NEED

Newspaper or thin card

Jam jar or 330ml (12fl oz) drinks can

Scissors

Garden twine (optional)

Seed tray

GETTING READY

Collect some paper or thin card – newspaper is best as it's flexible and easy to fold. Choose a container that will act as your 'mould'. The size of pot you create will depend on the mould you use. Jam and condiments jars come in a useful range of sizes, but a 330ml (12fl oz) drinks can works well. It needs to be rigid, so if it's a drinks can, use one that's still sealed, so it doesn't collapse as you use it.

1 Cut the paper or card into strips about 30cm (12in) long × 20cm (8in) wide, which is about the same size as an A4 piece of paper. Place the drinks can or jar on its side, at one end of the paper. Position it off-centre so the bottom lines up with one of the longer edges of the paper.

2 Roll up the can or jar in the paper. Fold in the long ends at the base in four neat overlapping folds; then sit the jar or can up and press it down onto the folds to squash them into place.

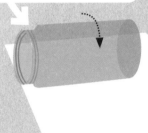

3 Gently pull out the can or jar with one hand, holding the paper roll with the other. Fold down the top 2cm (¾in) of paper into the pot, to hold the pot's shape. For newspaper, this is all it takes to keep the pot in shape. If you are using more rigid card, it is more likely to spring apart, so tie some twine around the middle to keep it in place.

4 Make a few more pots and nestle them together into a seed tray to help keep them in place, then fill with compost and get sowing.

GETTING TO GERMINATION

When you have your containers and seed packets lined up, and you know where you're going to house the planted containers, you're ready to sow. Seeds come in all forms, from large and chunky to speck-of-dust sized, but very few have special needs, and most will grow happily in the same medium.

POTTING COMPOST

The most widely available and best all-round growing medium for starting seeds is a peat-free seed compost. Seed compost has quite a light texture – seeds like a medium that isn't too heavy or lumpy, as they need air to start growing. Because they have inbuilt food stores, seeds don't need any additional enrichment in the soil until after they've germinated and grown some leaves. A bag of vermiculite, a natural mineral that is very light and absorbent, is also useful. It can be scattered over seeds that prefer to be covered very lightly once sown, and if a seed packet makes much of the fact that the contents need free-draining soil, you can mix a little in with the potting compost, too.

SOWING

Fill a container with potting compost to within about 4cm (1½in) of the top. Shake the seeds into your palm and either place or scatter over the surface of the compost. Place larger seeds between one and three to a small pot or cell, smaller ones a few more, and the tiniest may be planted 'by the pinch' rather than anything more specific. However small the seeds, try to distribute them as thinly and evenly as possible. Then cover to the depth suggested on the packet – this can vary from 1–2cm (½–1in) of compost, to the thinnest-imaginable scatter of vermiculite for those seeds that need light to germinate. Water lightly (a sprayer does the job best) and leave in a light, warm spot.

DATES AND LABELS

Don't forget to mark your pots with the name of the plant and the date the seeds were sown, even if you're sowing just a few different sorts. You may think you'll remember, but you won't. Use a label and an indelible marker so that the writing doesn't get washed away with repeated watering.

Write the sowing date on your seed label, so you'll know how long those particular seeds took to germinate.

DIFFICULT CUSTOMERS

You'll occasionally find a mention of 'stratification' on a seed packet. This indicates that these seeds need the conditions they'd go through in nature to prompt them out of dormancy — often a storage period at lower temperatures. If you come across the term, read and follow the directions carefully.

SOME SEEDS NEED A BURST OF COLD TO GET GOING

AFTER THEY'VE STARTED

After seeds are sown, the majority will germinate within two weeks, although some take just five or six days to sprout. It's always thrilling to watch the tiny shoots emerge, uncurl and start growing, but there are still a few stages to go through before your brand-new seedlings mature into young plants.

WATERING AND THINNING

New seedlings can stay in their starter pots for at least the first two weeks. Water lightly when the compost starts to feel dry to the touch. This may be as often as every day. The seedlings shouldn't dry out, but the compost must be moist, not saturated. Use tepid water and a watering can with a fine rose, or a hand sprayer.

If you planted several seeds to a pot and they all germinated, thin the seedlings as soon as there's a clear front-runner. Leaving the strongest (or the strongest two, if the pot is slightly larger), snip the stems of the others with scissors. Don't pull them out, or you'll risk disturbing the roots of the ones you want to keep.

The thinnings from vegetable seedlings such as carrots and beets are delicious – eat them as a gardener's treat.

LIGHT, FEEDING AND POTTING ON

Once a seedling has its first leaves, it starts to photosynthesize, and needs nutrition. Seed compost is sterile; it's good for germination, but doesn't contain any nourishment, so if your seedlings are going to stay in their original containers for a while, feed them a weak solution of liquid fertilizer once a week (make it around a third as strong as is recommended for adult plants).

Once germinated, plants need a good source of steady light, and this may mean turning pots regularly if they're on, say, a windowsill. If seedlings are outgrowing their original containers but it's too cold or too early to plant them in their final spots, they need potting on into larger pots. Fill the new pots with all-purpose peat-free compost (unlike seed compost, this contains some nutrients). Handle seedlings very gently, by the leaves rather than the stems, and try not to disturb the soil around the tiny roots as you move them from one pot to another.

HARDENING OFF

If your seedlings were raised indoors, harden them off before they're planted outside. You can do this by leaving them in a cold frame or greenhouse, or taking them outside during the day and bringing them indoors at night for a week or so. Remember that slugs and snails love tender seedlings, so raise the pots on bricks to make them less accessible.

CREATE A SEED CALENDAR

If it's your first year growing your own plants from seed, it's tempting to look forward to just as far as when your first plants will flower or crop. But take an overview of the whole growing year and you can make sure that your garden is productive from spring into autumn, and has something to offer, whether flowers or food, in most months.

KEEP A RECORD

Taking notes of what you planted when, plus when you expected to plant it out (and when you actually did), and when you expected it to flower or to crop (and when it actually delivered), is the biggest favour you can do for yourself when you're starting out. It helps you plan for future years and, if you're a novice, quickly adds to your gardening knowledge. You'll learn not just when things come into flower, but how long you can expect them to go on for; which flowers are short-lived, though gorgeous, and which bloom for weeks or even months (sweetpeas and cosmos are two standouts); and which crops are best grown little and often (lettuce, for example, unless you want 50 lettuces ready to eat in a single week), as opposed to which you should grow in a single sowing (tomatoes, and as early as possible, too).

It's best to have a dedicated diary for your record, and to fill it in once a week after seed-sowing starts. Make brief notes – you don't have to write at great length. Because whatever it says on the packet, and what friends and family may tell you, no-one's growing space or circumstances are quite the same as anyone else's, and the seeds that turn out to be standouts for you won't necessarily be the stars in someone else's garden. Growing is an intensely personal business.

SOWING AND GROWING AROUND THE YEAR

This calendar gives you an idea of the range of sowing/growing/flowering and cropping dates for just a few easy-to-grow flowers and vegetables. When you've got an idea of what you'd like to grow, work out your own version so that you can look forward to what is coming when, and decide which plants, if any, that you'd like to add to fill in any growing gaps.

FLOWERS

Sow Grow Flower

VEGETABLES

Sow Grow Harvest

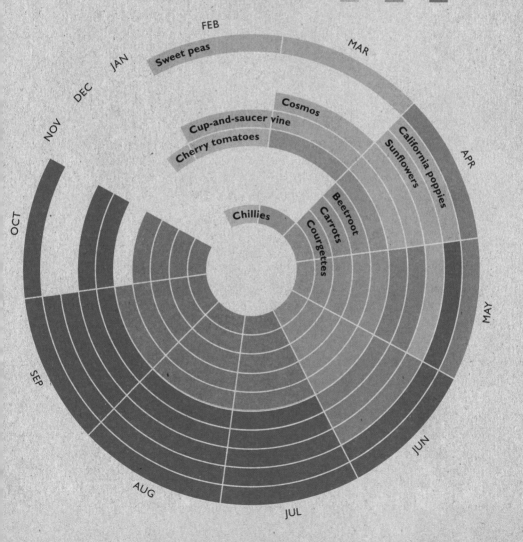

Sweet peas
Cosmos
Cup-and-saucer vine
Cherry tomatoes
Chillies
Beetroot
Carrots
Courgettes
Sunflowers
California poppies

JAN FEB MAR APR MAY JUN JUL AUG SEP OCT NOV DEC

CHAPTER TWO

FLOWERS FROM SEED

Many of the easiest and most rewarding stars of the flower garden are annuals – plants that germinate and flower in the same growing year. *Flowers from Seed* looks at different ways in which you can grow and enjoy some of the best, whether you're interested in creating a cutting garden to give you flowers indoors, looking at how you can support garden wildlife with your planting choices, or simply want ideas for planting up beds or containers. It introduces you to some biennials and perennials, too, defines hybrids, and explains the various growing patterns behind the labels.

The second part of the chapter contains 15 in-depth profiles of flowering plants that grow easily from seeds. Some are bound to be familiar – sweetpeas, foxgloves and sunflowers all feature, along with everything you need to know to give them a successful start, so that they can return the favour – while others (the Mexican cup-and-saucer vine, for example) may be new to you. What they all have in common is that they're straightforward to grow from seed, so whichever you choose you're set up for success.

USE YOUR SPACE

With such a vast range of seeds available, you should be able to find almost any plant you want to grow; getting the best effect with your plants in the space you have, though, is more a question of experience and judgement. Regimented planting rarely looks great – the most appealing spaces tend to be planted to look natural.

CAREFUL PLANNING

Of course, the 'natural' look isn't really anything of the sort; it's usually carefully planned. Look in any gardening magazine and you'll see artfully undulating drifts of flowers, higgledy-piggledy 'cottage garden' schemes that are full of different plants, but that nonetheless seem to have arrived at an even balance of colour, and so on. All use generous quantities of plants. The abundance you need for good planting wouldn't come cheap if you were buying adult plants, but it's easily arrived at when you're growing from seed.

Compact, colourful planting in a window box (above), contrasting with some light and airy 'natural' planting of tall-stemmed flowers (below).

There are a number of points that are good to remember, whether you're planting up a window box or a border.

- If you're planting in a limited space, vary the scale and include one or two outsized plants. Uniformly small plants in small spaces can look mean.

- When growing in containers, include some plants that will scramble over and downwards to soften the edges. It'll give a more natural effect.

- Plants tend to look better planted in uneven numbers – put your seedlings into groups of three, five and so on.

- Unless you're planting a vegetable garden or a cutting plot, avoid tight rows and the 'grown-on-a-grid' look; instead, think in curves.

- Allow for growth – plant your seedlings with the size of the adult plant in mind. It's good when planting looks generous, and when all the plants have grown to full size you don't want to see bare earth, but give everything enough elbow room.

- Compare flowering dates when you put plants together and calculate whether you want the flowers to appear sequentially – one plant arriving at its best as another starts to fade – or whether you want everything to be in full bloom in one go.

Remember, though, the real joy of gardening is being able to make your own space in your own style. Follow other people's good ideas when you come across them and think of plants as the ingredients in recipes that you can adapt to your own taste; in the end, the only person you have to please is you.

DON'T FORGET THAT CONTAINERS NEED REGULAR WATERING.

WHAT THE LABELS MEAN

If you're new to growing from seeds, the flowers you're most likely to choose first are the hardy annuals. They offer a wonderfully fast delivery on your investment – seeds planted in spring will be filling your garden with colour by midsummer, sometimes even earlier. When you're browsing the seed stands though, you'll come across half-hardy and tender annuals, too, as well as biennials and perennials.

FOXGLOVE

HOLLYHOCK

FLOWERING ON SCHEDULE

Annuals germinate, grow into adults, flower, set seed and die all in the same year, while biennials follow suit, but across a two-year timetable. In the first year, biennials germinate, and the young plants grow foliage and establish themselves; in the second, they flower, set seed and die, so if you choose a biennial, it will take a little longer to see a return. Finally, perennials last multiple years – the foliage dies back over winter, but the plant reappears in spring. Some perennials don't flower until their second year, others flower the year they're sown. And not all plants of the same type fall into the same group – both foxgloves (*Digitalis*) and hollyhocks (*Alcea rosea*), to take two very popular examples, have both biennial and perennial forms. It's not surprising that people become confused.

What's the takeaway if you're a novice? Always read the seed packet carefully – it will tell you which group this particular variety of your chosen plant belongs to, and then you'll know what to expect.

Finally, there's the difference between the 'hardy' and 'half-hardy' packet notes you'll see. It's straightforward – a hardy plant is one that will survive cold weather, including a frost, while a half-hardy one won't. You should wait until any danger of frost is past before you plant a half-hardy annual outside.

HYBRIDS

You'll also see 'F1' or 'F2' marked on seed packets after the plant's name. These tell you that the seeds inside are hybrids, made by crossing two different cultivars of a particular plant ('cultivar' is the abbreviation for 'cultivated variety') to create a new plant that combines the best of both parents' characteristics. F1 hybrids are direct crosses, while F2 are second-generation hybrids, made by crossing two F1 plants. Hybrids aren't any harder to grow, but you won't get the same uniformity or quality if you collect their seeds, because the seeds of hybrids don't breed true and some are completely sterile. Plant breeders put a lot of work into their development, so hybrid seeds tend to be more expensive.

GROWING FOR CUTTING

If you love having fresh flowers around the house, consider starting your own cutting garden — a patch grown purely to supply cut flowers. You don't need a huge amount of space; an area of just 2m² (21 ½ft²), carefully planned, is large enough to supply cut flowers for several months.

WORKING IT OUT

When you're growing exclusively for cut flowers, the layout of the bed is geared purely to productivity, as it is when you're planning a vegetable plot. Instead of mixing plants up, you sow or plant in tightly spaced rows. The idea is to maximize the number of flowers — the more you grow, the more you can cut. Two square metres is quite a small space, so choose just three or four varieties, and prepare the ground very well,

BISHOP'S FLOWER

SCABIOUS

digging in some extra manure or compost and raking the soil into fine drills. You should be able to reach all the flowers from the edge of the plot. Start your seed selection under cover and plant outside in April, when the ground starts to warm up.

MAKING A CHOICE

Everyone will have their favourite annuals, but most people find sweet peas essential in a cutting bed. You'll need a support for them to scramble up, and they may take as much as a third of your space, but with their fabulous scent and long growing season, they're a real luxury as cut flowers, and they have good vase life, too.

CUTTING-GARDEN STANDOUTS

These four cutting-garden standouts are all:

- easy to grow from seed
- have a long flowering season
- have an ability to mix happily with others in the vase

BISHOP'S FLOWER (*Ammi majus*)

- Tall annual with white, umbrella-like flowers that give an airy elegance to arrangements. Will flower for three months, between June and the end of August.

SUNFLOWER (*Helianthus*)

- Pick one or more of the many smaller, multi-flowered varieties and they will produce blooms on multiple stems from June all the way to early October.
- Try 'Vanilla Ice', with cream-coloured flowerheads, or 'Ruby Sunset', which has rusty-red flowers.

SWEET PEA (*Lathyrus odoratus*)

- The must-have of any cutting garden, with sweet-scented and astonishingly prolific flowers – at the top of their season they grow almost faster than you can pick them.

SCABIOUS (*Scabiosa*)

- Appealingly frilly flowers on long stems, in a range of pastel colours or purples. Try 'Fata Morgana', a subtle, pinkish apricot, or 'Black Knight', for a dark magenta-purple.

SWEET PEA

SWEET PEA AND LADY'S MANTLE

SUNFLOWER VANILLA ICE

MAKE A RUSTIC WIGWAM

Sowing the seeds of climbing flowers, or climbing vegetables, is just the start of their journey – they need something to clamber up to fulfil their potential. A support made by hand has a pretty and rustic appeal, and you can make it as high or as short as you need to suit your choice of plants. You'll need sturdy poles to make the uprights, and thinner, more flexible stems for the horizontal cross-ties. Hazel and willow are ideal for both.

YOU WILL NEED

5–7 hazel or willow poles at least 2cm (¾in) in diameter

10–12 flexible hazel or willow stems 1cm (½in) thick or less

Garden twine

GETTING READY

Ask your neighbours for any prunings they may have spare. If you can't find anything suitable locally, hazel and willow sticks are easy to find from online suppliers and can be bought in very affordable bundles. If you use willow, make sure it's well-dried, or it will sprout and start to grow when in the ground.

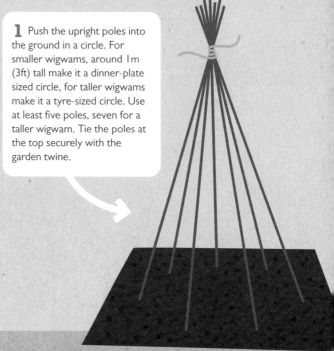

1 Push the upright poles into the ground in a circle. For smaller wigwams, around 1m (3ft) tall make it a dinner-plate sized circle, for taller wigwams make it a tyre-sized circle. Use at least five poles, seven for a taller wigwam. Tie the poles at the top securely with the garden twine.

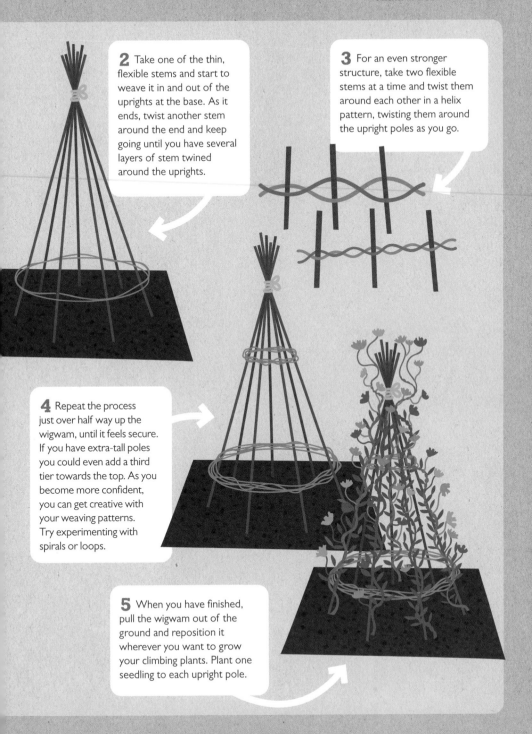

2 Take one of the thin, flexible stems and start to weave it in and out of the uprights at the base. As it ends, twist another stem around the end and keep going until you have several layers of stem twined around the uprights.

3 For an even stronger structure, take two flexible stems at a time and twist them around each other in a helix pattern, twisting them around the upright poles as you go.

4 Repeat the process just over half way up the wigwam, until it feels secure. If you have extra-tall poles you could even add a third tier towards the top. As you become more confident, you can get creative with your weaving patterns. Try experimenting with spirals or loops.

5 When you have finished, pull the wigwam out of the ground and reposition it wherever you want to grow your climbing plants. Plant one seedling to each upright pole.

GROWING FOR WILDLIFE

At a time when ecologists warn that global biodiversity is dramatically on the wane, even tiny growing spaces can act as feeding stations and refuges for wildlife. Simply choosing the right plants can help insects in particular, ranging from the familiar and photogenic – bumblebees and butterflies – to many other, less readily identifiable species, which nonetheless have crucial parts to play in our ecosystems.

POLLINATOR-FRIENDLY PLANTING

If you can make a mini-habitat in which insects will thrive, you'll also be benefitting creatures higher up the food chain, such as amphibians and birds. Begin by picking pollinator-friendly seeds. Annuals that would probably be on your to-grow list anyway are also popular with insects: the

cosmos (*Cosmos bipinnatus*) and honeywort (*Cerinthe major*) are just two examples (see pages 64–65 and 60–61). Look for seed packets with 'bee-' or 'pollinator-friendly' branding, or try the suggestions shown below.

Make a dedicated wildlife patch with four or five insect-friendly plants. Sow your seeds, and when the seedlings are large enough, plant them out together. This can be in a space as small as a window box or large container, or in a bed within a larger garden – you need to scale the number of each plant up or down accordingly. They should be planted quite densely, then left alone. Deadhead, so new flowers will continue to show up to supply pollen and nectar, but otherwise, leave it to the invertebrates and watch who turns up.

TEASEL

STANDOUT PLANTS FOR INSECTS

These four standout plants for insects are all:

- easy to grow from seed
- sure-fire hits with a wide range of species
- unfussy additions to any growing plan

BORAGE (*Borago officianalis*)

- A plant that may be familiar to you from the herb garden, with its deep-blue flowers and furry leaves. In folklore, it is called 'bee bread' – but its appeal isn't limited to bees.

TEASEL (*Dipsacus fullonum*)

- A biennial wildflower with spiky, mauve flower heads, loved by seed-eating birds as well as insects. It stays looking good even when the heads are reduced to winter skeletons.

OXEYE DAISY (*Leucanthemum vulgare*)

- Classic, large, white daisy flowers with a central yellow 'eye'; a perennial that seems to be a particular favourite with hoverflies.

ALYSSUM (*Lobularia maritima*)

- With dozens of tiny flower heads in shades from creams to pinks and purples, alyssum is not only a good filler, fitting neatly into small growing gaps, but is also a magnet for all kinds of insects.

BORAGE

ALYSSUM   OXEYE DAISY

GROWING FLOWERS WITH KIDS

When you enjoy hands-on gardening – being outside, making plans, and watching the plants cooperate and the plans work out – it's natural to want to pass your interest on to the next generation. And provided that you catch them young enough and can offer something that doesn't call for too much patience when it comes to waiting for results, a lot of children love a gardening project.

GREEN FINGERS

Get them while they're young: the under-10s are more likely to join in enthusiastically if they're offered a project than, say, the over-12s. If children have good memories of times spent in the garden, they're more likely to return to it when they're older – you may be helping to create the gardeners of the future. Here are two projects that quite small children will usually enjoy and manage with only a little help.

FLOWERS YOU CAN EAT

Edible flowers seem just sufficiently unusual to pique junior interest. Set up a large container or a small corner of a bed so that it's ready to plant a selection. Oversee from a distance, just to ensure the seed goes in at more or less the right depth, and that watering happens. The four easiest options from seed are wild pansy (*Viola tricolor*), violet (*Viola odorata*), marigold (*Calendula officinalis*) and nasturtium (*Tropaeolum majus*); all flower generously. Once they're in bloom, encourage the young gardener to pick off the flowers for a vivid bowl of salad.

VIOLET

NASTURTIUM

PANSY

MARIGOLD

SUNFLOWER HEIGHT CHART

It's an old one but a classic: the 'who grew the tallest sunflower?' contest. Shop around for the tallest variety of sunflower you can find. 'Giraffe' and 'Mongolian Giant' can both grow over 4.25m (14ft) tall, with flowers as much as 45cm (18in) across. Oversee the sowing and subsequent watering (see pages 70–71), and the necessary staking and support as the plant grows tall, but try not to take over – remember whose project this is.

Once the plants start to grow in earnest, organize a weekly measuring ceremony (as they get taller, measuring will become an increasing logistical challenge). Children can use the stems as height charts to measure their own height, too (just mark the stem with an indelible marker). And finally, when the sunflower starts to wither, make the most of the flower, either by letting the seeds dry out before showing your child how to roast them with a little salt, or by hanging the whole head in a corner of the garden as an all-natural feeding station for the birds.

PLANT A TEXTURED & AIRY CONTAINER

From a few tiny seeds, it's possible to create textured container displays with huge impact. Mixes of airy annuals will make a beautiful focal point next to a door, or in the midst of smaller potted plants. Flowering plants that are tall but not too bushy are ideal, and ornamental grasses will add movement and texture.

YOU WILL NEED

Seeds from one of the combinations below

Seed trays and small pots

Large container, at least 40cm (16in) in diameter at the top

Peat-free compost

Pea sticks or a metal hoop support

GETTING READY

The beauty of this project is in the planning and design. Choose one of the three designs below. Each design contains complimentary plants that vary in height and texture, including one with a feathery or grassy foliage to give the container an airy, dreamy look. You'll need two of each plant, so sow at least three in case one fails.

DESIGNS

Hot colours

Annual persicaria
  (Persicaria orientalis)
Zinnia 'Benary's Giant'
Pheasant tail grass
  (Anemanthele lessoniana)

Pretty pastels

Cloud grass
  (Agrostis nebulosa)
Annual scabious
  (Scabiosa atropurpurea)
Cosmos 'Seashells'

Meadow effect

Foxtail barley
  (Hordeum jubatum)
Cornflower
  (Centaurea cyanus)
Baby's breath
  (Gypsophila paniculata)

1 Sow the seeds in February or March indoors.

2 When the seedlings are large enough, lift them and plant into the small pots. In mid-May, gradually acclimatize the seedlings to the outdoors, by moving them outside during the day but bringing them in again at night.

3 When all risk of frost has passed, plant two of each plant, six in total, into your container, and water in well. Position the plant supports.

4 Place in a sunny position with all-round light, to prevent the tall stems leaning. Water every other day in summer, but check the moisture in the pot before you do. Peat-free compost can remain moist for longer than expected – scrape away the surface to make sure it is dry about 5cm (2in) down before watering.

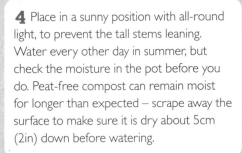

Hollyhock *Alcea rosea*

Lazily waving hollyhocks are a quintessential summer sight. They thrive in baked, stony soil, happily seeding themselves along gravelly roadsides, or at the base of south-facing walls. The towering spires of flowers reach heights of 3m (10ft) or more, as either double forms – which form bright pom-poms – or open-flowered single forms, which benefit pollinators.

SOWING

Sow the seeds in seed trays filled with seed compost, indoors, anytime from February to May. Space out each flat, papery seed on a seed tray. Sprinkle 2mm (1/16in) of compost or grit over the top to keep the seeds in place. Water from below, by floating the tray in a shallow bowl of water until it seeps up from the base to saturate the top. Place the tray in a propagator, or cover with a polythene bag to keep warm.

GROWING AND CARE

Seedlings take up to two weeks to germinate. When they are large enough to handle, gently prick out and plant them into larger pots. Move them outside in May, and keep them well-watered in a sunny position until they have bulked out into small, leafy plants.

PLANTING AND PICKING

Hollyhocks are better suited to growing in the ground than in containers. Plant them towards the back of borders, making sure they receive sunlight. They are tough but short-lived perennials, which form a rosette of flowers in their first year and go on to flower the following summer.

Hollyhocks will self-seed around gardens, but if seedlings appear at the front of borders, dig them up and pot them until they have bulked up, then re-plant them at the back of the border.

THE BALLOON-LIKE BUDS HAVE A BEAUTY OF THEIR OWN, BEFORE THEY BURST INTO FLOWER. THIS IS A SINGLE-FLOWERED FORM.

DOUBLE FORM

THREE TO TRY

'Black Knight' – striking, near-black, silky-looking single flowers on a tall plant

'Chater's Mix' – a cottage garden favourite; double flowers in a mix of sumptuously rich shades

'Halo White' – large, white, single flowers with a lemon yellow centre, which glows in the sunshine

Columbine *Aquilegia*

Columbines (*Aquilegia*) are cottage garden favourites with designer good looks, flowering in late spring. Each flower is a masterwork of architecture, with shapely spurs or whirling fractal petals. Its nodding form also makes it one of the daintiest perennial flowers, often known as 'Granny's bonnet'. Flowers can be a single colour, or have an inner and outer set of petals of different shades.

SOWING

Sow the seeds outdoors in early summer or autumn. They can take a while to germinate and often need a period of chilling, so it's best to sow them in a seed tray or pot rather than directly into your garden soil. Fill the tray with seed compost, and water until it is thoroughly damp. Sprinkle the seeds evenly over the surface. They are tiny, so only need a sprinkling of vermiculite to keep them in place.

GROWING AND CARE

If you sow in autumn, leave the tray in a sheltered position outdoors where it won't be disturbed, such as in a cold frame. If you have to leave it out in the open, put chicken wire over the top to prevent cats or birds digging the surface. Germination will begin in the spring.

If you sow in early summer, you may not see germination for several weeks. Try putting the seed tray in the fridge for two weeks, then place it back outside, to help speed up germination.

PLANTING AND PICKING

Columbines prefer to grow in fertile garden soil, in sun or part-shade, but can be planted in containers if they are kept well-watered. Though perennial, columbines are quite short-lived, but once you have them, new seedlings will pop up everywhere, usually in different colours and forms to the parents, which can be part of their appeal.

FLOWERS WITH INNER AND OUTER PETALS IN DIFFERENT HUES ARE KNOWN AS BI-COLOURED.

THREE TO TRY

Aquilegia chrysantha – tall, lemon-yellow flowered species with upward-facing flowers that have long, elegant spurs

'Ruby Port' – a deliciously deep, wine-red cultivar with pom-pom flowers that are packed with petals

'Swan Lavender' – a good form for pots, with large flowers and delicate lavender-and-cream colouring

Snapdragon *Antirrhinum*

These bright flowers are great fun for kids – you can snap the blooms open and shut like little puppet dragons. They are happy either in the ground or in large containers, flowering from June to September. A short-lived perennial, they will come through winter only if grown in a very sheltered spot.

SOWING

Sow indoors from January to March. Snapdragons need plenty of warmth to germinate. Fill a seed tray with seed compost, then water, before sprinkling over the small seeds evenly. Cover with a very thin layer of compost or vermiculite. Put the tray in a propagator, or seal inside a polythene bag and place on a warm, bright windowsill.

GROWING AND CARE

The seeds take anything from a week to 21 days to sprout. When they are large enough to handle by the leaves, gently prick them out and pot them into small, individual containers. Grow them on in cooler conditions, but keep them indoors until early May, when you can start to acclimatize them by moving them outside during the day and back inside at night for a week or so.

PLANTING AND PICKING

Only plant snapdragons outside when the risk of frost has passed. Though their flowers grow in spikes, snapdragons can form quite bushy plants, so give them about 30cm (12in) of surrounding space if planted in the ground. In pots, they will be happy to be nestled in a little more closely.

Deadhead the flower spikes as soon as the last blooms fade to keep the plants blooming all summer. They make good cut flowers, but make sure you cut the flowers just as the lowest few flowers have opened.

TRIM BACK THE BUSHY SPIKES AS THE FINAL FEW FLOWERS BLOOM.

THREE TO TRY

'Purple Twist' – a very tall cultivar with white flowers and painterly purple stripes

'Madame Butterfly' – frilled, double flowers in shades of salmon pink, red and white make this a sumptuous choice

'Night and Day' – dark-green foliage sets off the striking velvety flowers of deep red and pure white

Cornflower *Centaurea cyanus*

This native wildflower once speckled crop fields throughout the UK with electric blue flowers, bursting into life wherever the soil was disturbed. They are tall and upright, with ruffled pin-wheel blooms, and are a magnet for bees and butterflies. Grow them in full sun, either in the ground or in a container.

SOWING

Cornflowers are best sown in the soil where you want them to flower. Give them plenty of space and sow several in a drift for impact. Sow in spring for flowers later in summer. If you sow in autumn, the hardy seedlings are tough enough to survive most winters and will flower earlier the following year.

Weed the soil and rake it to a fine, crumbly texture. Draw curving or wave-shaped drills in the soil about 12mm (½in) deep. The seeds are just large enough to pick up with your fingers and have a distinctive tuft. Scatter them along the drills a few centimetres apart and cover them over with more soil, then water lightly. To grow in containers, sow the seeds in trays or small pots.

GROWING AND CARE

Seedlings take up to two weeks to germinate. Thin out those growing in the ground to about 30cm (12in) apart, or plant pot-grown seedlings into the soil with the same spacing. Cornflowers prefer poor, dry soil, so don't feed the soil or pots where they are planted.

PLANTING AND PICKING

Cornflowers are tall plants and can flop without support. Some pea sticks pushed in amongst the plants will help. They will keep flowering for months if you deadhead or pick them regularly. Cut the flowers as they open to just above a leaf node, and pop them in a vase. The remaining stem will produce more flowers.

PACKED WITH BOTH POLLEN AND NECTAR, CORNFLOWERS ARE A BOON FOR BEES.

THREE TO TRY

'**Black Ball**' — striking flowers of a deep, almost chocolatey, burgundy hue; great teamed with ornamental grasses

'**Blue Boy**' — perhaps the bluest of the electric blue cornflowers, tall and strong-growing

'**Polka Dot Mixed**' — a rich mix of jewel-coloured flowers in pink, mauve, deep purple, and blue

Honeywort *Cerinthe major*

Honeywort (*Cerinthe major*) is an otherworldly plant with unusual deep-purple, pendulous flowers surrounded by clustering bracts of almost iridescent green-tinged mauve, sometimes with white speckles. It's a reliable and easy to grow plant, doing well both in the ground or in pots. A freely draining soil is best and although it will tolerate some shade, the colours truly shine in full sun.

SOWING

Sow indoors from February to April, or directly outdoors until June. For earlier flowers, you can also sow outside in September and October. As long as the soil is free-draining and the position sunny, the young plants are hardy enough to come through most winters.

Indoors, sow seeds in a tray, with two seeds in each cell, and cover with a sprinkling of compost or vermiculite, then water lightly. To sow outside, weed and rake the ground and sow into shallow drills, then re-cover, and water.

GROWING AND CARE

Early spring sowings need gentle heat, so place the seed tray in a propagator or inside a polythene bag, and keep on a warm, bright windowsill. Outdoors, the seeds only need the occasional watering if it's dry. Germination can take anything between 5 to 21 days.

PLANTING AND PICKING

When the risk of frost has passed, and if they are big enough to handle, plant indoor-grown seedlings to their final position in May. One of the benefits of autumn-sowing is that honeywort makes a stylish partner for tulips and narcissi.

Honeywort is a magnet for bees and other pollinating insects. The seedheads pop open when ripe and will gently self-sow, so once it's happy in your garden, it's probably there to stay.

YELLOW HONEYWORT IS A RARE BUT NO LESS BEAUTIFUL FORM.

OTHERS TO TRY

Borage (*Borago officinalis*) – this close relative of *Cerinthe major* has starry flowers and is a pollinator magnet

Comfrey (*Symphytum uplandicum*) – tubular flowers reminiscent of *Cerinthe*, and useful for making homemade, liquid plant-feed

Purple sage (*Salvia officinalis* 'Purpurascens') – a sun-loving herb with similar, purple-tinged foliage, that's also edible

Cup-and-saucer vine *Cobaea scandens*

Don't let the exotic looks of this climbing plant put you off – it is not tricky to grow. It germinates easily and will take off quickly in a sunny spot, scrambling over any support using its large, twining leaves, ideal for covering a bare fence or archway. Large, goblet-shaped flowers appear in late summer.

SOWING

Soak the papery seeds in water for an hour before you sow to help them stick to the compost and germinate more quickly. Sow from February to April indoors, in individual pots filled with seed compost. Push one or two seeds into the compost on their sides and water well.

GROWING AND CARE

Put in a propagator or seal the pots inside a polythene bag to keep them warm, and place on a warm, bright windowsill. Remove coverings once the seedlings start to get tall. They will be several centimetres high before you know it, so letting them have their own pot with a small stick to cling to makes their energetic growth easier to control. If you don't have much room indoors, sow later, in April, so you can move them outside before they become rampant.

PLANTING AND PICKING

Once the risk of frost has passed, young cup-and-saucer-vines can be planted out in the ground or large containers. Spend a week acclimatising young plants by moving them outside in the day and back in at night.

Cup-and-saucer vines are perennial plants from Mexico, but are mostly grown as annuals in cool temperate climates. But in very mild areas, in towns or on the coast, they may survive the winter and keep growing.

They can be late to flower, but full sun and shelter help speed things up.

SEEDLINGS TAKE OFF QUICKLY, SO MAKE SURE YOU HAVE A SUITABLE PLACE TO KEEP THEM BEFORE PLANTING OUTSIDE.

OTHERS TO TRY

Cobaea scandens f. alba – the white form of the species with creamy, sometimes almost greenish blooms

Rhodochiton atrosanguineus – exotic annual climber with deep, purple, bell-shaped flowers that are smaller than *Cobaea*

Morning glory *(Ipomea tricolor)* – tender climbing annual with striking blue flowers that open with the morning sun

Cosmos *Cosmos bipinnatus*

Easy to sow and low-maintenance to grow, cosmos quickly make a big impact with their vibrant pink, white or candy-striped blooms. Useful for plugging gaps in summer borders, they grow equally well in containers. Cosmos thrive in most garden soils as long as it isn't waterlogged or very dry, and prefer full sun.

SOWING

Sow from late March to late April, for flowers from early July. Germinate under cover, either on a sunny windowsill or in a greenhouse. The large, long seeds are easy to handle, so this is an ideal seed to sow with children.

Plant the seeds in a seed tray, two to each cell, or in small, 9cm- (3½in-) wide pots with two seeds to each pot. Fill to the top with seed compost, press down firmly then scatter the seeds on top. Cover with a thin layer of compost or vermiculite about 3mm (⅛in) deep and water lightly.

GROWING AND CARE

As long as they have warmth, cosmos seeds germinate in as little as one week. If both seeds in a pot or cell germinate, gently prick out the smallest one, transplanting it into another pot if you want to avoid waste.

Plant young plants outside or into larger pot displays after the risk of frost has passed, normally in late May. Before this time, seedlings in trays may outgrow their homes (check to see if the roots are poking out of the drainage holes), and will need to be planted into a slightly larger pot before they are planted again into their final home.

PLANTING AND PICKING

Cosmos look particularly lovely when planted in groups of five or more. With regular deadheading, they will keep blooming until late October, or the first frosts.

Cosmos also make excellent cut flowers, lasting 7–10 days. Cut stems just as the buds are opening.

THE PORCELAIN-FINE FLOWERS GLOW WHEN BACKLIT BY THE SUN.

THREE TO TRY

'**Seashells**' – fluted, almost cylindrical petals make this an unusual and intriguing flower form

'**Rubenza**' – a shorter cultivar, ideal for containers, with the deepest pink-red, velvety petals

'**Purity**' – classic old, white cultivar, tall and fresh, adding airy grace to flowerbeds

Foxglove *Digitalis purpurea*

With imposing spires of flowers, foxgloves have real impact. They are biennial, which means they form a clump of leaves in their first year and flower in their second. Foxgloves sometimes act as short-lived perennials, flowering for a year or two afterwards.

SOWING

Tiny foxglove seeds can be as small as specks of dust, and can cause illness if swallowed, so handle with care. Sow between June and August. Simply scatter the seeds over open, raked soil in your garden and leave them to grow or, if you want to position them more carefully, sow in a seed tray first. Fill it with seed compost, press down the surface, then place it in a tray of water until the moisture saturates through to the top. Then evenly sprinkle the foxgloves seeds on the surface. They need light to germinate, so don't add more soil or compost on top.

GROWING AND CARE

Germination takes 14 to 21 days. Prick out seedlings in trays when big enough to handle and plant them in individual pots. Grow the plants in a partly shaded place, out of drying sunshine or wind, and keep them well-watered.

PLANTING AND PICKING

When they have filled out their pots, plant them into your garden, preferably in autumn, but spring planting is also fine. The plants will send up tall spires of flowers in midsummer. Just as the first spike has faded, cut it off above the rosette of leaves and more, smaller, side flower spikes will form, giving a second flush of flowers.

TAP THE DRIED FLOWER
HEADS TO RELEASE THE
TINY SEEDS.

THREE TO TRY

'Pam's Split' – unusual, split petals make this a fascinating foxglove, which is shorter than most

'Alba' – a white form with glowing white spires, excellent in shade

'Summer King' – known as the strawberry foxglove, with salmon-pink flowers mottled with a snakeskin pattern

California poppy *Eschscholzia californica*

The classic California poppy is a silky, tangerine-orange flower that positively glows in the summer sun, but many cultivars flower in delicately dusky colours with ruffled, petticoat-like petals. The flowers twist into closed cones at night and open in sunlight into little bowls of brightness. They die away after summer, but often self-seed year after year.

SOWING

Sow directly outside in April. California poppies grow well in most soils, but especially poor, dry soils. They are short plants, so they'll be most visible planted in front of others. This is also the sunniest spot, which will encourage the most flowers.

Weed and lightly rake the soil to a fine texture, then water gently. Poppy seeds are tiny, so scatter them slowly and evenly, taking care not to sow them in a clump. Sowing them in a sweeping motion gives a natural effect. The tiny seeds need light, so don't cover them. Sow more seeds every two weeks until June, for a succession of flowers.

GROWING AND CARE

Seedlings take 14 to 28 days to appear – watch out for their fern-like leaves. Remove any competing weed seedlings that may creep in. Once they are a few centimetres high, thin the seedlings to about 15cm (6in) apart. You can replant the thinnings, but take care not to damage the taproot.

PLANTING AND PICKING

Each flower is quite delicate and short-lived, but a swathe of these plants will bring a succession of blooms over a few weeks. As cut flowers they will only last a couple of days, but make pretty posies.

CALIFORNIA POPPIES ARE A GOOD PLANT CHOICE FOR THE EDGE OF A GRAVELLY PATH.

THREE TO TRY

'**Peach Sorbet**' – beautiful selection
with full, double, silky flowers in a gorgeous
shade of soft peach

'**Ivory Castle**' – elegant white flowers
with an ivory-yellow tinted centre and
blue-green foliage

'**Misson Bells**' – slightly taller than average,
with ruffled petals in a mix of fruity shades

Sunflower *Helianthus annuus*

Famed for towering, record-breaking flower heights, sunflowers also have surprising variety in colour and form. The largest are sunny yellow, but smaller forms have red, orange, burgundy, rich brown or creamy-coloured blooms. If you let some of the flower heads ripen at the end of the summer, small birds will be attracted to the nutritious, oil-rich seeds.

SOWING

Sow sunflower seeds from mid-April to early May. You can sow them directly into soil that's been weeded and raked. Young seedlings are a favourite of slugs and snails, so it may be worth sowing and growing the young plants in pots first until they reach about 30cm (1ft) tall, before planting them into the ground. To sow in pots, fill with seed compost and make a 2cm (¾in) deep hole in each one with your finger or a pencil. Drop in two seeds and close over the compost.

GROWING AND CARE

Germination takes two to three weeks. Protect seedlings germinating in the open soil with a sprinkling of organic slug pellets or a barrier. For seedlings in pots, remove the weakest if both seeds sprout. If the roots don't get too damaged, you can try replanting the seedling in another pot. The seedlings grow quickly, so keep them well-watered.

PLANTING AND PICKING

Plant potted plants outside from late May. Push a stake into the ground alongside tall types. Plant in a sunny position, but bear in mind that sunflowers turn their heads to face the sun, so avoid planting them in a north-facing bed, where they may spend their time pointing away from you to the south.

Branching sunflowers produce many more stems than the single-stemmed, tall forms, so are best to grow if you want plenty of blooms for picking.

ORANGE-RED CULTIVARS SUCH AS 'EARTHWALKER' ADD A TOUCH OF GLAMOUR.

THREE TO TRY

'Velvet Queen' – a tall but branching sunflower with burnished red petals and chocolate brown centre

'Magic Roundabout' – a multi-stemmed cultivar ideal for cutting, with many-shaded pastel blooms

'Titan' – the one to grow for a sky-scraping flower, reaching over 3m (10ft) high

Sweet pea *Lathyrus odoratus*

Some sweet peas are worth growing for their heavenly scent alone, but with the bonus of frilly petals in myriad hues, they are a must-have for lovers of flouncy flowers. They are fast-growing, annual climbers, looking lovely grown on a wigwam of sticks, and the large seeds are easy to handle, making this an ideal seed to grow with small children.

SOWING

Sweet peas are hardy, but better off sown indoors to protect them from hungry mice or wet soil conditions, in which they can rot. They'll need a cool but bright place, such as an unheated greenhouse, or next to a garage or shed window. Sow them in either October to November, or late January to April.

Fill trays or pots with seed compost and make holes about 2cm (0.7in) deep in each. Toilet-roll tubes make excellent sweet pea pots – they allow space for the plant's long tap root and can be planted whole, the cardboard quickly disintegrating into the soil. When sown, water well and place in a cool, bright place.

GROWING AND CARE

Avoid putting sweet peas in a warm spot as the seedlings will grow too quickly, outgrowing their pots before you are ready to plant them outside. If they do start to shoot up, push sticks (old chopsticks work well) into their pots for them to cling to, which will stop them becoming tangled together. Pinch out the growing tips to encourage more stems.

PLANTING AND PICKING

Plant out the seedlings in April, taking care not to disturb the roots too much, and water in well. These hungry plants will appreciate a feed every few weeks.

Pick the flowers frequently to encourage the plants to produce more, and remove any seed pods to see. As soon as sweet peas set seed, they stop flowering.

PINCH OUT THE GROWING TIPS WHEN SEEDLINGS ARE ABOUT THIS SIZE.

THREE TO TRY

'**Almost Black**' – intensely scented, with very deep purple, velvety flowers, ideal for picking

'**Turquoise Lagoon**' – iridescent colour, flowers begin mauve and age to turquoise blue, the result of cutting-edge breeding

'**Cupani**' – classic cultivar more than 300 years old, with highly scented pink and purple blooms

Tobacco plant *Nicotiana*

Tobacco plants are slender plants with trumpet-shaped flowers that release a sweet scent in the evening. They are popular as a patio plant, where they perfume the air on warm evenings. The shorter, brightly coloured forms are jaunty enough to enliven a container, while some taller, white-flowered forms have an elegant, designer-garden quality that looks graceful in a flowerbed.

SOWING

Sow from late February to April. Tobacco plants are half-hardy annuals, which means they need some warmth to get going. Fill a modular seed tray with seed compost and sprinkle a very small pinch of the tiny seeds into each cell. Don't cover the seeds with compost – leave them exposed to light. Water the tray by floating it in a shallow bowl of water, or wet the compost before you sow, to avoid disturbing the seeds on top.

GROWING AND CARE

Put the seed tray in a propagator, or inside a polythene bag, making sure the plastic doesn't touch the surface (tent it up with a couple of short sticks if necessary). Place on a bright windowsill in the warmest room you have for three weeks – germination takes about 7 to 20 days. When the seedlings are large enough to handle, plant them on into larger pots and move them to a cooler position.

PLANTING AND PICKING

Tobacco plants can be planted outside, in full sun or part shade, once the risk of frost has passed. Young plants need a period of acclimatization before planting out; move them outdoors during the day for gradually longer periods over a week, before planting into their final position.

Most tobacco plants make excellent cut flowers, lasting for several days in a vase, but take care when handling as the stems can have a sticky residue.

THE NIGHT-SCENTED, WHITE TOBACCO PLANTS, IN PARTICULAR, WILL DRAW NIGHT-FLYING MOTHS AND BATS.

THREE TO TRY

Nicotiana sylvestris – very tall form towering to 1.5m (5ft), topped with elegant, creamy-white trumpets

'Tinkerbell' – a dainty cultivar with unusual terracotta flowers, with a greenish reverse side

Nicotiana alata 'Grandiflora' – known as the jasmine tobacco plant, it's one of the most highly scented forms

Love-in-a-Mist *Nigella damascena*

Never was a flower so aptly named – the soft, blue flowers of love-in-a-mist are swathed in fine, wispy foliage, and bloom in delicately swirling colours of blue, white and mauve. This cottage-garden favourite is easy to grow, with the flowers followed by attractive, sculptural seed heads. Once growing in your garden it will often self-seed around year after year.

SOWING

Sow in April for flowers from July. For a longer flowering period, make more sowings every two weeks until early summer. As a hardy annual, you can also sow in early autumn – the seedlings will survive the winter to flower the following May or June.

The seeds can be sown in the ground or in a seed tray. Make sure the ground is free of weeds and raked to a crumbly texture. Draw shallow drills in the soil (make them wavy or curved for a more natural pattern), scatter seeds along them and cover them over with a little soil. To sow in a seed tray, fill it with seed compost, sprinkle the seeds thinly over the surface and cover with a thin layer of compost.

GROWING AND CARE

Seedlings take between two to four weeks to appear, quickly developing characteristic wispy, ferny leaves. Thin out clustered seedlings in the soil to about 15cm (6in) apart. For seedlings grown in trays, gently prick them out when they are large enough to handle and grow them on in small pots.

PLANTING AND PICKING

Pot-grown plants can be planted outside when the roots have filled the pot. Plant them in groups to create a dreamy effect. Cut flowers will last about a week in a vase, but leave some flowers to develop into the attractive seedheads.

THE LIFE CYCLE OF LOVE-IN-A-MIST, FROM TINY SEED TO POPPABLE POD.

THREE TO TRY

'Miss Jekyll White' – ethereal white flower with a deep indigo eye and grey-green foliage

'Mulberry Rose' – a cottagey cultivar with a mix of bright pink and creamy-white blooms

'Delft Blue' – the blue-and-white washed flowers look painterly, with striking, deep-red-topped seedheads

Nasturtium *Tropaeolum*

This tender annual has unusual round leaves and exotic but cheerful-looking flowers in shades of sunny yellow, orange and red. The variety of forms is wide, with romping, trailing cultivars which, given some support, can scale a fence, to dainty plants that will be happy in hanging baskets. The leaves, flowers and fresh seeds are also edible, and add a peppery taste to salads.

SOWING

Nasturtiums need warm conditions to germinate, so sow the seeds indoors in March or April. Alternatively wait until May, June or July to sow outside. Fill a seed tray or small pots with seed compost, make a 1–2cm (⅜–¾in) deep hole in each and drop two seeds in. Cover them over with compost and water well. For the larger, climbing type of nasturtium, try sowing the seeds in the ground alongside a fence or obelisk, where they can climb.

GROWING AND CARE

Nasturtiums usually germinate within two weeks, but can be slower to germinate outdoors. If both seeds in each cell germinate, pinch out the smallest one, leaving the other to grow on. When the seedlings have reached about 10cm (4in) high, you can plant them, but take care to acclimatize them to the outdoors if they have been growing indoors. Wait until mid to late May to harden them off for a week, moving them outdoors in the day and back indoors at night, gradually increasing the time spent outside until the risk of frost has passed.

PLANTING AND PICKING

Plant nasturtiums in a sunny position. They love heat, so a sheltered spot in front of a south-facing wall or fence is ideal. They thrive on poor soil, so don't feed them. They begin flowering in June or July depending on when sown, and will continue blooming until the first autumn frost.

NASTURTIUM FLOWERS ARE A COLOURFUL WAY TO SPICE UP A SALAD.

THREE TO TRY

'Ladybird Rose' – a dwarf variety, ideal for containers, with unusual, dusky-pink flowers with hints of lemon yellow

'Red Troika' – a trailing variety with deep-scarlet flowers and interesting speckled, variegated foliage

'Milkmaid' – very pale-yellow, almost white flowers, blooming on a mid-sized plant

Zinnia *Zinnia elegans*

Popping with neon-bright colour, zinnias are a visual treat. They have hot, bright jewel tones, and flamboyant, velvety pom-pom flowers in late summer. The largest rival dahlias in form and colour, but without the fuss of the winter care. Zinnias are perennial, but grown as annuals in colder climates. They love the heat, so will perform better in sunnier summers.

SOWING

Zinnias dislike root disturbance, so sow directly outdoors into seed drills in late April or May, in warm, weed-free and raked soil. If the spring is a cold one, or if you want earlier flowers, sow them indoors in individual pots from March to May. Cardboard or paper pots are a good option as they minimize root disturbance (see pages 28-29).

Zinnia seeds are just large enough to handle individually. Put two or three in each pot and cover with a light sprinkling of compost or grit. Put in a propagator, or cover with a polythene bag and place on a warm, bright windowsill.

GROWING AND CARE

Germination takes seven to ten days. Water with fresh tap water only and take the propagator lid or bag off for a while every other day, to allow fresh air in, which will help avoid damping-off disease.

PLANTING AND PICKING

Plant out pot-grown seedlings after the risk of frost has passed, taking care not to disturb the roots. Grow them through netting, or use sticks pushed into the soil around the plants to stop them flopping. Zinnias can suffer from fungal disease in cold, rainy summers. In sunny years, though, they thrive if they are given plenty to drink. Avoid watering the leaves and aim to drench the roots.

Zinnias make excellent cut flowers, with a vase-life of seven days or longer.

TAKE CARE NOT TO DAMAGE THE ROOTS OR DELICATE STEMS WHEN YOU REMOVE THE SEEDLINGS FOR PLANTING.

THREE TO TRY

'Giant Purple Prince' – spectacularly huge zinnia with hot pink flowers that rocket upwards

'Envy' – a pale lime green, dahlia-like zinnia loved by flower arrangers

'Zahara Double Mixed' – a cultivar with resistance to cool, wet weather, in vivid mixed colours

CHAPTER THREE

FOOD
FROM SEED

If previous experiments in growing your own food
have been limited to a pot of mint and one or two tomato
plants, this chapter will show you that you can be a lot
more ambitious – and succeed. Of course, there's
nothing wrong with a single, sun-warmed tomato
picked straight from the plant, but pair it with a just-
harvested corn cob and add some lettuce that was still
in the ground an hour ago, and you have the makings
of a truly delicious lunch. Not only that, but you'll know
that your meal's food 'miles' have been limited to the
few metres between garden and kitchen, and that the
food you're eating is free from chemical residues and
hasn't been pumped up with intensive fertilizers.

Food from Seed takes you through the basics of growing
vegetables from seed without assuming you have a lot of
space at your disposal. You'll find features on fast-growing
herbs and microgreens, plus profiles of a range of crops,
from the familiar – tomatoes and carrots – to some you might
not have thought of – pak choi and cucamelons. You'll
discover how to harvest a trugful of vegetables, rather
than a handful. And once you get growing, the taste
alone will probably be enough to get you hooked.

USE YOUR SPACE

Even if your ground space is limited, you can probably grow more than you think. You can use windowsills and window boxes, and plant in both small and large containers, and grow bags, too. And if you do have open soil to grow in, carefully choosing which varieties to raise in it will ensure you get the very best out of the space you have.

BED OR CONTAINER?

Unless you've got a large garden with masses of open ground for growing, ground space that can be dedicated to vegetables tends to be quite precious. If you have just one or two beds, use big containers and grow bags for any crops that will cope with them, saving your open ground for those vegetables that either prefer it, or that grow too big to manage in a pot. The latter include sweetcorn, the full-sized climbing peas and beans, and some of the larger pumpkin and squash varieties.

Beetroots, tomatoes, chard, courgettes and small carrots, called finger carrots, to name just a few examples, will generally grow well in a pot provided that it's large enough. Look through seed catalogues for 'patio', 'baby' and 'miniature' varieties – this indicates a smaller variety that's often been especially developed to grow in a more confined space. The crop from these cultivars will often be smaller, with a more delicate flavour, and may mature faster than the larger varieties.

Match your crops to your space: there's an edible plant for every corner of your garden, whether you're growing in containers, raised beds, or open ground.

WHICH VEGETABLES SHOULD YOU MULTI-BATCH?

Most salad greens, spinach, French beans and carrots all grow and crop comparatively fast, and it's best to sow several small batches, at two- or three-week intervals, to lengthen the cropping season, give you a steady supply, and avoid a huge harvest ripening all at once, creating a glut.

Tomatoes, chillies, peppers and aubergines are much slower croppers and, in northern European climates, will need a full season to mature: sow just once, early in the season.

CARROTS

FRENCH BEANS

SPINACH

SHORT-TERM CROPS

If you're going to sow successional crops – that is – stagger the sowing of batches of salad greens, for example, across spring and summer, or plant any quick-maturing crop into space that will be taken up by another crop later in the season, it's usually easier to put them in the ground. After you've harvested the first crop, you can rake, add some compost, and sow the fresh crop in place more easily in open ground. If successional crops are in containers, you may have to empty them completely between crops, which is a more labour-intensive option.

GET YOUR TIMING RIGHT

Once you know what your space represents in what-you-can-grow terms, and you've thought about which vegetables you want to grow, make a month-by-month plan. This should take account of which vegetables will crop only once a season, and which more than once. The plan you make is the first step to keeping a growing record which, in turn, will be invaluable for future seasons' planning.

DO THE MATHS

Make your plan using the information on the seed packets you've bought. If, for example, you've bought carrots that you're planning to grow in a container, look at the sowing date, how long germination is expected to take, and how long the estimated time is between germination and cropping. Copy them down, starting in the month you'll sow and carrying through to the month you'll expect to crop, adding where the crop is going to grow. Repeat with each packet, putting the information onto a chart or in a notebook. When you've finished, you'll have the basics of a monthly 'to do' list.

REAL-LIFE GROWING

This initial plan is what you hope will happen – but while some aspects of it will be under your control, others won't. Every growing space is individual. How much sun, for how long each day, what the soil is like (when you're growing in open ground), which way the plot faces and how sheltered it is, will all play a part in your success. Then there's the weather – a very hot/cold, wet/dry spring or summer will affect how well your crops grow.

With these factors taken into account, your first growing year will inevitably be a learning curve. Whichever way things go, keep a record of each crop against your original monthly plan. Make notes of the date the seeds of each variety were sown, whether under cover or not, the dates they germinated, when the young plants went into the ground or a container, the dates you harvested various crops – plus how successful/plentiful/tasty they were. Note any problems you had with pests, whether or not you gave additional liquid feed, and what, if any, protection in the form of cloches or fleece you used. Next year, you'll be surprised by how absorbing all this information will be! You'll have the beginnings of a full growing record, including which varieties performed well, which didn't (indicating that you should try different cultivars next year), how real-life timings measured up against your initial projections, and which other factors seem to have had either positive or negative effects.

Noting problems as well as successes in your veg-growing story will help ensure that you can avoid them next year.

SOWING VEGETABLES

You've planned your space, worked out your timings, and you're ready to sow, either into the ground, or into pots or cells. Start with a sowing list; the broad rules of sowing are described on pages 26–31, but there are a few other things, such as staggered sowings of certain crops, to take into consideration.

Make a list of the sowing dates only from your month-by-month plan. If you originally envisaged a single mass-sowing event, a glance at your schedule will show you that it doesn't work that way. You'll probably find that you'll have several sowing dates across your seed selection, with some seeds to be started off indoors and some outside.

INDOORS/OUTDOORS

If you're a seed novice and the packet tells you that you have the option to either sow the seeds earlier 'under cover', or later, out of doors, if you have the space, opt for the indoors option. It gives you more control over the seeds' germination stage and means that by the time they go outside, they'll be sturdy little plants. After a few years of growing from seed you'll develop a confidence about sowing outside, but for first-timers, indoors feels safer.

Some packets will only give the outdoor option. Usually, they'll also give a later sowing date than the under-cover one. Follow it; don't be tempted to get the seeds in earlier to give them more time to grow – outdoor sowing is mostly about temperature, and until the ground starts to warm up a little, seeds are unlikely to germinate. Some crops, notably beetroots and carrots, don't mind being sown while temperatures are still very low outside, but they're the exception rather than the rule.

BE PREPARED

If you're impatient to get on, use the intervening time to prepare the ground thoroughly.

Use a garden fork to break up any big clods of earth in your sowing bed and pick out any big stones by hand. Dig in some well-rotted compost if you have some (if you don't, it's not essential; you can add enrichment to the soil at any time). Remove any visible weeds. Finally, use a rake to go over the whole bed several times, raking it in both directions and weeding out any roots or clumps of vegetation that are lurking below the surface as you go. By the time you've finished, the bed should look all ready to be raked into drills of fine earth and planted up.

Dig over your bed with a fork, then rake thoroughly, walking backwards, so you're not walking on the soil you've already raked.

ON THE SMALL SIDE: MICROGREENS

When you want a quick crop to add flavour, freshness and vitamins to a meal, even in midwinter, microgreens could be the answer. If as a child you ever grew a tray of mustard and cress, you already know the basics: these tiny crops are grown indoors and eaten as small seedlings. There are many varieties of seed that can be sown and raised as microgreens, all relatively speedy germinators.

EATS SHOOTS AND LEAVES

'Microgreens' describes how the seeds are grown, rather than which specific variety they are. Suitable seeds include a range of salads and herbs, among them spinach, various lettuce varieties, rocket, mizuna, radish, fennel and coriander. The best are fast to germinate – you won't want to wait for the small-though-delicious crop – and have a distinctive taste, even as small shoots.

Growing microgreens is a good way to use up leftover half-packets, but you'll also find packets sold for the purpose – they're generally good value, as they contain more seeds than regular packs.

MUSTARD

BEETROOT

RADISH

MICROGREEN STANDOUTS

BEETROOT
- Pretty, red-veined leaves; earthy and sweet. If you wait for an extra few days then pull up the whole plants, the roots have a definite 'beet' taste

ROCKET
- Warm and peppery, even the very young leaves have a slight crunch

FENNEL
- Feathery little greens with a slightly anise taste, milder than fully grown fennel, but distinctive – good in an omelette, or with fish

CORIANDER
- Baby coriander has the same rich, complex flavour as the adult herb, and can hold its own mixed in with milder leaves

ROCKET

CORIANDER

GROWING IN MINIATURE

To grow microgreens, you need a seed tray, some all-purpose compost, and seeds. If you have more than one sort, sow each in a different tray as they're unlikely to germinate at exactly the same rate. As they're grown indoors, you can start microgreens off at any time of year.

Fill the tray with compost and draw two or three drills along its length using a pen or your finger. Sow the seeds evenly but thickly along each drill. You should sow much more densely than you would a regular crop, as these seeds will only ever reach seedling size, so won't get too tightly packed. Cover seeds lightly with compost, water, and leave somewhere light and warm to germinate. Most microgreens will germinate within five to seven days, and be ready to harvest within two weeks. Leave the seedlings until they are around 5–7½cm (2–3in) high before harvesting (cutting with scissors).

GROWING FLAVOUR

Pots of herbs on a terrace or a windowsill are one of the simplest luxuries you can grow. Few of us will have space for them all, so the six below are an edited selection of the best options: all have qualities that give them appeal in the pot as well as the kitchen, smelling and looking good. Some will also draw in pollinators.

SIX EXCELLENT ALL-ROUNDERS

CHIVES
(*Allium schoenoprasum*)

Chives look pretty growing, and the buds, blue flowers and tubular leaves can all be eaten. They're also good for pollinators.

Sow under cover in March;
plant out in April; harvest from July.

DILL
(*Anethum graveolens*)

With feathery leaves and umbrella heads of yellow flowers, dill should be sown directly where it is to grow (it can reach a height of 120cm (47in). It's popular with pollinators, and its fronds have a subtle anise taste that goes with fish or summer salads.

Sow direct into container in April;
harvest from June onwards.

CHERVIL
(*Anthriscus cerefolium*)

CHIVES

A small-leafed plant that is happy in partial shade, chervil isn't often found in shops. It has a subtle liquorice flavour that works best with fish or omelettes, or as part of a mixed-leaf salad.

Sow direct into container in March;
harvest from May onwards.

DILL

CHERVIL

CORIANDER
(*Coriandrum sativum*)

Sow coriander direct into quite a deep container (it has long roots). It tends to bolt if it's grown in full sun; shift it to partial shade and the foliage will beef up. The pungent leaves can be cooked into spicy dishes or eaten fresh.

Sow direct into container in June; harvest from August onwards.

BASIL

CORIANDER

SORREL

SORREL (*Rumex acetosa*)

Sorrel is classed as both a herb and a salad leaf; it's a perennial, which will establish itself in the ground or in a (large) container. The big leaves can be picked by the handful and have a subtle lemony taste. It's uncommon in the shops, and well worth growing.

Sow under cover from February; plant out in April; harvest from June.

BASIL (*Ocimum basilicum*)

A stalwart in Mediterranean cooking, basil needs plenty of sun. Pinch out the growing tips to encourage a bushy plant. To harvest, pull off the top leaves first. You can grow many varieties from seed, including Thai basil, which has a pronouncedly spicy flavour, and purple basil, which is more strongly aromatic than the green variety.

Sow under cover from February; plant out in May; harvest from July.

GROWING WITH KIDS

How much children can grow for themselves – from seed to crop – will depend on what age they are. If they spend time with you while you're starting your vegetable season, though, they'll quickly make the connection between what you grow and what you eat which, at a time when food production can seem remote from everyday life, is a good lesson in itself. And they're usually happy to eat what they've been involved in growing, too.

PICK YOUR OWN

Encourage smaller children to choose some of their own seeds while you're ordering, ideally ones with a built-in appeal: cherry tomatoes and sugarsnap peas that can be picked off the vine like sweets, or radishes they can harvest themselves. When the seeds arrive, involve them in the planting, the potting up, and subsequent watering. Eventually, go on a pick-your-own journey, sampling everything you've grown.

GROW A RECIPE

With slightly older children, choose a recipe – ratatouille, tomato sauce or sweetcorn salsa are all contenders – and agree that you'll grow all (or at least most) of the ingredients, then cook it together. For a salsa, for example, you might need sweetcorn cobs, tomatoes, spring onions and coriander. Make it a joint enterprise, but get them to do at least some of it themselves.

HAPPY HALLOWEEN

Grow a pumpkin for Halloween – it's a project with a nice clear aim and end date. If you have enough space to grow a plant each, hold a competition for the best result.

Plant under cover in April, with each single seed, placed on its side, in a small pot. Keep somewhere warm and light, and the young plant or plants can go outside at the end of May. A pumpkin needs sun and very fertile soil – dig in a few spades of compost or manure, and mound the soil slightly before planting the seedling at the centre. Protect with a cloche overnight while the plant is small, and water regularly around the edge of the mound – pumpkins are thirsty, but direct watering may rot the plant. When the fruits start to develop, feed every week with liquid fertilizer, and prop the young pumpkins away from the earth with a piece of slate or tile (this discourages rot).

Children of 10 years and up can usually manage all of this, either as a group activity or, with an occasional reminder, by themselves. The pumpkins should be mature and ready to harvest by mid-October – then all you need to do is to carve them!

GROW A POTAGER IN A POT

Potager-style gardens are ornamental vegetable plots, with a style of planting that aims to be beautiful as well as productive, rivalling flower gardens in colour and form. You can make a mini-potager vegetable garden in a pot, or really a large container, which will overflow with crops that look as good as they taste.

YOU WILL NEED

Seeds from one of the designs below

Small pots or seed trays

Peat-free compost

Well-rotted farmyard manure

Large container
(see Step 2 for details)

GETTING READY

Using plants that will be happy companions is key to making a mixed vegetable container work.

Each of the designs below contains plants that will grow alongside each other without too much competition. With different habits, they will grow in different sections of the space, either trailing over the edge of the container, filling out the middle or climbing above.

DESIGNS

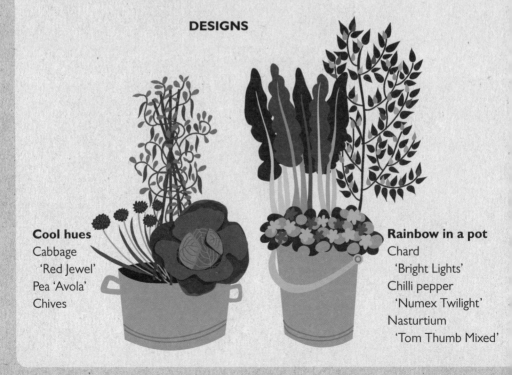

Cool hues
Cabbage
  'Red Jewel'
Pea 'Avola'
Chives

Rainbow in a pot
Chard
  'Bright Lights'
Chilli pepper
  'Numex Twilight'
Nasturtium
  'Tom Thumb Mixed'

1 Sow your crops according to the seed packet instructions, in seed trays or small pots, indoors in February or March. Grow the seedlings in pots until the risk of frost has passed. Grow a few of each plant, just in case some fail.

2 Edible crops normally need plenty of room to perform at their best, so a container can be a cramped environment. Choose one as large and deep as you can find – an old tin bath or large flexi-trug would both work well. Drill or hammer in some drainage holes if needed.

3 Half-fill the container with the compost. Most edible crops are nutrient-hungry so to give them a boost, top up the container with the farmyard manure and mix in well, but leave a gap of a few centimetres below the rim to allow for watering.

4 Position your plants. Put the tallest – this will be the pea or the chilli – at the back of the pot, with canes or sticks for support. Place the chunkier cabbage or chard in the centre, and the lower growing plants – the nasturtium or the chives – towards the front and side, so they can trail over the edge of the pot.

5 Plant them and water well. Feed every two weeks with a balanced liquid fertilizer through summer. The fruiting crops – the peas, chillies, nasturtium flowers and chive leaves – can be picked as they grow, but harvest the bigger cabbage or chard in late summer, when it has reached full size.

Beetroot *Beta vulgaris*

Beetroots are great all-rounders: sweet, earthy and delicious. Harvest the roots when they're small and tender, or leave them longer for larger beetroots. Use the iron-rich leaves, too, steamed or in salads. The roots come in multiple colours – deep red, in pink-and-white candy stripes, or with orange or white flesh.

SOWING

Sow the seeds from April onwards. Batches sown every couple of weeks will yield beetroots into autumn. Plant the seeds under cover in individual cells, or directly into the ground as the soil starts to warm up. Planted under cover, germination takes seven to ten days; sown outside, seeds should germinate within three weeks.

Beetroot seeds come out of the packet in feathery little clusters; sow them into cells filled with seed compost, placing one cluster in each. Cover with a thin layer of compost or vermiculite. Water sparingly.

Beetroot plants like a free-draining soil, with plenty of organic matter. If you're sowing outside, rake the soil, mark out rows (drills) 30cm (1ft) apart and sow at 20cm (8in) intervals. Cover lightly with soil, then water.

GROWING AND CARE

Several seedlings usually germinate from each seed cluster. To harvest single beetroots, snip all but one stem off with scissors (pulling them up disrupts the root of the remaining seedling). If you want groups of smaller beetroots, leave alone and treat as a single plant.

To transfer seedlings outdoors, squeeze the cell so that the soil comes out in one piece, and pop the soil ball into the bed, firming lightly around it.

Water the beet plants regularly in dry weather.

HARVESTING

Beetroots are ready to harvest between six and twelve weeks from germination, depending on the weather and the size you prefer to eat them. Pull up one plant after six weeks; if the root is smaller than you'd like, wait a couple of weeks before trying again.

FANCY, GOLDEN OR STRIPY BEET VARIETIES ARE OFTEN EXPENSIVE TO BUY, BUT THEY'RE EASY TO GROW FROM SEED.

THREE TO TRY

'Barbabietola di Chioggia' – pretty pink-and-white striped roots with a mild, earthy flavour; good raw in salads

'Boltardy' – classic, deep-red, medium-sized beetroot, very sweet

'Touchstone Gold' – bright, golden-yellow flesh; delicious roasted

Chard *Beta vulgaris* subsp. *cicla*

With vivid, multicoloured stems and ribs, and dark-green leaves, chard is both a pretty and a delicious crop – choose from colours including white, yellow, pink and red. Pick the vitamin-packed stems and leaves young and eat them raw, or wait until the plants get bigger and braise or stir-fry them.

SOWING

Chard is best sown directly where you want to grow it, in a spot that gets plenty of sun, whether that's in a container or a bed. It's decorative enough to make an effective partner with midsummer flowers and looks good in a mixed flower-and-veg container. It's also tough, so you can sow it outside from late March; repeat sowings will give you a crop all the way through to winter. Unlike some other greens such as spinach, chard doesn't usually bolt or run to seed, and in a mild winter, the plants are likely to survive to give you fresh leaves in spring.

Chard likes moist soil with plenty of organic matter. Sow the knobbly little seeds into prepared ground. If you're putting them in a bed, sow thinly into pre-raked drills about 45cm (18in) apart, and cover them over with 1cm (½in) of soil. In a container, it's easier to scatter the seeds sparingly on the surface, cover over with a thin layer of seed compost, then water.

GROWING AND CARE

Germination time varies depending on the temperature of the soil; in lower temperatures, it may take as long as three weeks, but if the soil has already warmed up a little, seeds may start to sprout after only a week or ten days. If necessary, thin the seedlings after two or three weeks.

TENDER YOUNG CHARD LEAVES
ARE DELICIOUS EATEN RAW
OR LIGHTLY STEAMED.

HARVESTING

Leaves and stems should be ready to pick around ten weeks after germination. Cut single stems from the base of the stalk. New leaves will quickly replace them.

THREE TO TRY

'Bright Lights' – a good spectrum of coloured stems – red, orange, yellow and white – from a single variety

'Charlotte' – tender red stalks and red-veined leaves; relatively small and grows more tidily than some larger varieties

'Fordhook Giant' – a taller variety with elegant, long white stems

Broccoli *Brassica oleracea* Italica Group

Broccoli is often used to refer both to the plant known as calabrese, which has green, cauliflower-like heads and thick stems, and to sprouting broccoli, which has narrower, leafier stems and smaller green, purple or white heads. You can harvest one or the other over summer and autumn and, overwintered, into the following spring.

SOWING

Start off both calabrese and sprouting broccoli under cover, in small pots or individual cells filled with seed compost, planted around 1cm (½ in) deep, or sow direct into the ground from April, after any chance of frost has passed. Indoors, allow two seeds to a cell. Germination takes up to two weeks; thin to one seedling ten days later. Plant out when the seedlings reach 10cm (4in). Shake gently until the soil ball comes out whole, then plant into the prepared bed.

Broccoli likes rich, moist soil, so a bed with some organic matter dug in is best. Rake drills, leaving about 45cm (18in) between them, then sow two or three seeds together, leaving 30cm (12in) between the groups; thin out to a single seedling when large enough to handle, and drape mesh over the baby plants to protect the seedlings from birds or caterpillars.

GROWING AND CARE

Calabrese is grown for same-year harvesting, but sprouting broccoli is tough enough to overwinter and then harvest in the early spring. So the growing plan for the two types is different – as well as being sown in spring and harvested in summer, sprouting broccoli can be sown outside as late as August and harvested in January or February.

YOU CAN EAT EVERY PART OF PURPLE SPROUTING BROCCOLI – LEAVES, STEM AND HEAD.

HARVESTING

Calabrese started in spring should be ready to harvest from August onwards; cut the central head and the plant will continue to produce new shoots and heads. Harvest the spears of sprouting broccoli when they have plenty of side leaves and the flower heads are formed, but before they open.

THREE TO TRY

'**Burbank**' – vigorous, sprouting broccoli with multiple white spears and many side shoots

'**Green Magic**' – a calabrese variety that produces lots of tasty side shoots and a denser central head

'**Purple Sprouting Early**' – sturdy option that you can overwinter and harvest in early spring

Pak choi *Brassica rapa* Chinensis Group

With a tender, crunchy core and flavourful green or pink outer leaves, pak choi is unfussy to grow. Use it in stir-fries or salads, or just steam it until it wilts and serve with a little soy sauce. It copes happily in semi-shade, but has a slight tendency to bolt (shoot up and run to seed) if grown in full sun.

SOWING

Pak choi can be started under cover or sown where it is to grow, but baby plants are susceptible to slugs and snails, so under cover is usually best, whether in a propagator, a cold frame, or even on a windowsill. Sow in May, into small pots or cells full of potting compost, three or four seeds to a pot (you can stagger sowings every couple of weeks through to July for repeat cropping). Cover to a depth of about 2cm (¾in), and water. Seeds germinate within seven to ten days. Thin out the seedlings to two per pot when they reach 5cm (2in).

GROWING AND CARE

Pak choi can be grown in a container or grow bag, or in beds. It prefers moist soil with plenty of organic matter, so dig in some well-rotted manure before planting. In a bed, plants should be put in at 30cm (12in) intervals, in drills 45cm (18in) apart; and given similar space in a container or grow bag. It's a thirsty crop, so keep it well-watered.

HARVESTING

You can harvest the outer leaves of pak choi as a 'cut-and-come-again' crop, cutting off the outer leaves of the younger plants as you need them, or leave the plant to develop a core and harvest the whole thing. It usually takes between four to six weeks to grow large enough to harvest the leaves, and another two or three to grow to a full-size head with a heart. If you stagger sowings, you can be harvesting pak choi into October.

IF YOU TAKE JUST THE OUTER LEAVES OF A PAK CHOI PLANT, THE INNER CORE WILL CONTINUE TO GROW.

THREE TO TRY

'Canton White Dwarf' – short-stemmed variety, particularly good for harvesting baby leaves

'Joi Choi' F1 – tough, elegant variety with deep green leaves and white stems

'Red Choi' F1 – leaves turn a deep red as the heads ripen

Chilli *Capsicum*

Many hundreds of different chillies are available. They are close relatives to sweet and bell peppers, which have similar needs and are also worth a try. All need a long season in the sun to ensure ripe fruit, so it's best to start chillies early. They'll crop most reliably under some sort of cover – whether a greenhouse or a cold frame – but will be happy outdoors in a hot summer.

SOWING

Sow the seeds in February or at the beginning of March, and grow the plants in containers, so they can be moved around to get the maximum amount of sun, or to shelter them, as necessary.

Start them off in individual small pots full of seed compost, a couple of seeds to a pot, and cover over lightly with compost or vermiculite. Keep them warm in a propagator if you have one, or improvise by enclosing the pot loosely in a plastic bag and placing it in a warm, light place such as a sunny windowsill. The seeds should germinate within ten days.

GROWING AND CARE

If more than one seed germinates, thin out after a week, leaving the strongest shoot. Pot up as necessary as the seedling grows, watering regularly and ensuring that it doesn't outgrow the pot to the point at which roots are growing through the drainage holes. Once the chilli plant has grown to around 30cm (12in), pinch out the central shoot at the top of the stem with your finger and thumb; this encourages bushy side growth to develop. When the plant begins to flower, start a weekly feed of liquid potash fertilizer.

HARVESTING

The fruit will develop from late July onwards. If you don't like too much heat, pick the chillies while they're still young – the longer a chilli stays on the plant, the darker its colour will be, and the hotter its flavour.

HOMEGROWN CHILLIES
NEED AN EARLY START AND
PLENTY OF SUN.

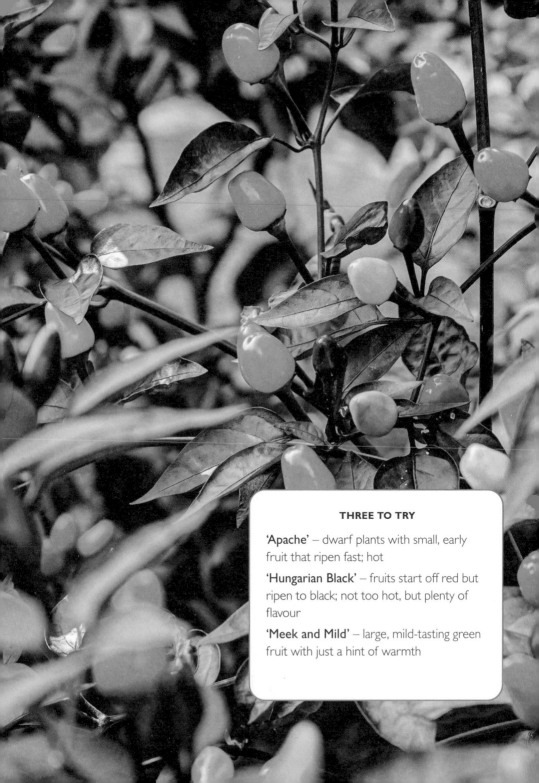

THREE TO TRY

'Apache' – dwarf plants with small, early fruit that ripen fast; hot

'Hungarian Black' – fruits start off red but ripen to black; not too hot, but plenty of flavour

'Meek and Mild' – large, mild-tasting green fruit with just a hint of warmth

Cucumber *Cucumis sativus*

Homegrown cucumbers divide into those that must be grown under cover, in a greenhouse, and those – sometimes called ridge cucumbers – that can be grown outdoors. There are many more varieties available as seed than you'll ever see in the shops, so it's well worth raising a few of your own.

SOWING

Both indoor and outdoor cucumbers need a long growing season. Sow seeds under cover in late February or early March. Use small pots or cells filled with potting compost, two of the flat seeds planted on their sides in each pot, then water, and place somewhere warm and light – a propagator or a windowsill. Germination should happen in seven to ten days. Thin and pot individual seedlings into larger pots when they have grown 10cm (4in) tall. Once the young plants have three pairs of leaves, pinch out the growing tips.

GROWING AND CARE

The seedlings are ready to go into their final growing spot, whether a pot in the greenhouse or an outside bed, by early June. Cucumbers are scramblers – they need staking, and sometimes extra support with string or netting. Fill containers – which should be at least 30cm (12in) diameter – with a rich compost including plenty of organic material. Outdoors, choose the sunniest location you have and dig some compost into the bed before settling the young plants in, leaving 45cm (18in) space around each. Water regularly around the plant's base, not over the cucumber itself. As the plants start to flower, a cucumber will develop behind each female flower. In plants under cover only, a flower with no fruit behind it should be pinched out – it's a male, and if it fertilizes a female flower, the resulting cucumber will taste bitter.

CUCUMBER SEEDLINGS

HARVESTING

When the fruit is at least 15cm (6in) long, cut it off the plant with a knife. Fresh cucumbers have the best flavour when they're not too large.

THREE TO TRY

'Marketmore' – old-fashioned outdoor variety, dark-green fruits with slightly bumpy skin

'Mini Munch' F1 – lots of small, crisp, thin-skinned fruits on a neat, compact, outdoor plant

'Tokyo Slicer' F1 – outdoor variety with slim, long, lightly ridged fruits

Courgette *Cucurbita pepo*

Courgettes are easy to grow from seed and are extraordinarily prolific croppers – allow just one plant per person and you should have plenty of courgettes all season. The fruits can be long, pattypan- or globe-shaped, and come in white, green, yellow and stripes. The gorgeous yellow flowers can be eaten, too.

SOWING

Courgettes are best started under cover; if you plant the seeds outdoors where they're to grow, they'll need some protection. In April, indoors, sow into cells, or small individual pots of seed compost, laying one seed in each cell sideways rather than on end, and covering with around 1cm (½in) of compost. Outdoors, start in May when the soil has begun to warm up, and sow two or three seeds in a group into a bed or container. Courgettes like rich, moist soil, so dig some compost in place before planting. Cover them for protection (you can use a cloche, but a large, cut plastic bottle, see page 23, will do just as well) and leave the covering on until two weeks after germination.

GROWING AND CARE

Courgette seeds germinate within ten days indoors or two weeks outside. Outside, thin groups to a single seedling after two weeks and remove the cover. Seedlings started under cover should be hardened off for a week before planting out. Courgettes drink a lot, so water generously around the base of the plant, trying not to get the leaves wet (this can cause mildew). Give a weekly feed of organic liquid potash fertilizer, too.

HARVESTING

To anyone new to them, the big surprise is just how many courgettes each plant produces – you'll be picking several every few days. Harvest while they're still small and tender, about 10cm (4in), to keep the plant cropping. Pick small courgettes with the flowers still attached – the latter are good in salads, or stuffed and fried.

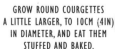

GROW ROUND COURGETTES
A LITTLE LARGER, TO 10CM (4IN)
IN DIAMETER, AND EAT THEM
STUFFED AND BAKED.

THREE TO TRY

'Eclipse' F1 – round, striped green fruit; good to cook stuffed

'One Ball' F1 – round, yellow fruit with a mild flavour; delicious eaten raw

'Zucchini' F1 – classic long, deep-green courgette; generous cropper

Squash *Cucurbita maxima*

Squashes can look very different from one another – knobbly or smooth textured, orange, green, striped, even an almost-grey. Essentially, though, there are two types: summer squashes, which should be eaten as soon as they're harvested, and winter squashes, which crop later, have thicker skins, and can be stored. Grow several to get the best of both worlds.

SOWING

Sow the seeds in late February or early March. The seedlings are very susceptible to cold, so it's easiest to start them off under cover. Using small pots or cells filled with potting compost, plant two of the flat seeds on their sides in each pot, cover lightly with compost, and water. Place somewhere warm and light, such as a propagator or a windowsill. Seeds should germinate in seven to ten days. Thin to the stronger seedling and pot on when the seedlings have grown 10cm (4in) tall. Harden the young plants off for a few days before planting them outside in late May.

GROWING AND CARE

Squashes prefer a sheltered spot, with plenty of sun and regular watering. Dig a bucket's worth of homemade compost or well-rotted manure directly around where you're going to plant, then dig in your seedling and add an additional mulch of compost. Even smaller varieties should be given a surrounding space of 90cm (3ft) to spread into. If planting in large containers or grow bags, use the same rich soil mix and put just one plant in each. Put a small stick alongside the seedling, so you can locate the middle of the plant for watering after it reaches its full sprawl. The fruit develops behind the female flower. As it grows, keep it clear of damp ground (which can cause rot) by adding a layer of straw underneath or propping it up with bricks.

HARVESTING

Summer squashes, such as the pattypan varieties, can't be stored. They are tenderest and have the best flavour harvested like courgettes, while still quite small. Winter squashes, such as butternut types, are ready to harvest when they feel firm to the touch and their stems have begun to dry out. Cut through the stem, leaving as long a stalk as possible.

THE FRUITS OF CLIMBING SQUASHES MAY NEED ADDITIONAL PROPPING OR SUPPORT AS THEY GROW.

THREE TO TRY

'**Crown Prince**' – winter squash: beautiful, blue-green skin; great for storing

'**Vegetable Spaghetti**' – summer squash: when cooked, the flesh separates into dozens of tender, spaghetti-like strands

'**Winter Hercules**' – winter squash: a neat plant that yields very large, butternut-type fruits

Carrot *Daucus carota*

Carrots that are homegrown and eaten within hours of harvesting have a flavour far better than any you can buy. You can grow them in beds or large containers and, planted in batches, you can harvest them all the way through from June until late autumn.

SOWING

Carrots like fine, free-draining soil. Once germinated they don't like being moved, so the seeds should be sown directly, either in rows into very well-prepared and raked beds, or in large, deep – at least 75cm (30in) – containers.

Seeds can be sown from April onwards as the ground starts to warm up. Scatter the small seeds as finely as you can into pre-raked drills set around 30cm (12in) apart, or onto the surface of soil in containers, then cover lightly and water thoroughly. They should germinate within 21 days.

GROWING AND CARE

Provided that the soil they're growing in is fine enough (heavy soil can split the roots), and they are watered regularly in dry weather, carrots don't need a lot of care. Carrot root fly, though, is a common pest that you need to guard against. It lays its eggs in the soil around the carrot plants and the larvae burrow down to eat holes in the growing carrots. The fly is attracted to the carrot scent, which is given off most strongly when you're thinning the seedlings. The best protections are first to sow the seed finely, to minimize the need to thin and second, to cover the young seedlings with a layer of light fleece, so the fly can't reach the soil. Early carrot crops are more susceptible than later ones.

HARVESTING

Carrots should be ready to harvest 12 weeks after germination; if some thinning of seedlings is needed before this, you can eat the seedlings, which are tiny but tasty.

HARVEST CARROTS ON THE SMALL SIDE FOR THE SWEETEST FLAVOUR.

THREE TO TRY

'Flyaway' F1 — bred to be resistant to the pesky carrot fly; great first choice for the novice grower

'Parmex' — compact and rounded shape; good selection for growing in containers

'Purple Haze' F1 — exotic-looking, purple-skinned carrot with a mild, sweet flavour; delicious eaten raw

Tomato *Lycopersicon esculentum*

Homegrown tomatoes crop generously and, depending on which varieties you choose, can be used for everything from salads to passata. They're easy to grow provided that you get them started early, so the fruit has time to ripen. Grow several different kinds and you can enjoy a multicoloured harvest in a range of shapes and colours.

SOWING

Choose varieties carefully according to what space you have. Generally, bush tomatoes are easier to manage if you're new to growing or short on space. Some cordon types can need a lot of staking, but they are great for greenhouse growing, where supports are less likely to blow over. A greenhouse will also get you a longer cropping season, whatever type you choose. Tomatoes can be grown successfully in containers and grow bags, as well as in beds.

Sow the seeds under cover at the end of January or in early February. Fill small pots or individual cells with seed compost, then scatter two or three of the light, flat seeds onto the surface of each pot or cell, and cover them with a very thin layer of compost or vermiculite. Water, and leave somewhere warm and light (in a propagator if you have one, or on a windowsill). Seeds should germinate in around a week.

GROWING AND CARE

Thin the seedlings if necessary as they grow, and pot the strongest up singly into larger pots. Harden them off before planting outside in late April or May. If you're using containers, a full-sized bush tomato will need quite a large pot – 30cm (12in) diameter – while a grow bag will fit three plants. After flowering starts, give an organic liquid potash feed once a week, and thin out the leaves to allow the developing fruit to get plenty of sun.

HARVESTING

Pick individual tomatoes or whole trusses as soon as they ripen; any green fruit at the end of the season can make pickles or chutney.

DWARF TOMATO VARIETIES GIVE A GOOD CROP AND TAKE UP VERY LITTLE SPACE.

THREE TO TRY

'Gardeners' Delight' – heavy cropper with tasty, medium-sized fruit; comes in bush or cordon types

'Losetto' – red cherry tomato, good for hanging baskets as it grows in dense 'cascades'; good blight resistance

'Sungold' – yellow cherry tomato; early trusses of tiny fruit with a delicious sweet flavour

Cucamelon *Melothria scabra*

Cucamelon, or 'mouse melon', is a small, perennial climber hailing from Mexico. In colder, northern climates it's grown as an annual and, given enough sun, will produce a crop of olive-sized fruits that look just like minute watermelons. They taste like cucumber with a faint flavour of lime – and have a very high cute factor.

SOWING

Cucamelon hates the cold, but grows rapidly when it gets going, so sow the seeds under cover towards the end of April, in small pots or cells filled with seed compost. Cucamelons are generous croppers, so one or two should be plenty for most households. Place the pip-shaped seeds blunt-end down, two or three to a pot, then cover with 1cm (½in) of compost or vermiculite and leave somewhere warm, such as a propagator or a sunny windowsill. Be patient: cucamelons can be slow to germinate – up to three weeks isn't unusual.

GROWING AND CARE

Thin out as necessary, potting up single seedlings into individual pots two weeks after germination. They can go outside four weeks later, provided that the weather is warm. Move into their final containers (pots should be at least 25cm/10in in diameter) when the seedlings are 20cm (8in) tall. Cucamelon is a vine, so needs to be given something to climb up – a container alongside a trellis in a sunny spot is perfect for it; failing that, give it some stakes to climb up and it will quickly attach itself with tendrils. It's a scrambling climber, but planted somewhere warm and sheltered it can grow as tall as 2.5m (8ft).

HARVESTING

Cucamelon fruits appear behind the tiny yellow flowers and will be ripe from late July onwards. Pick them while they're still firm for the best flavour; left too long on the vine they'll go woolly and take on a bitter taste.

CUCAMELONS CAN BE SURPRISINGLY HEAVY CROPPERS. SNACK ON THEM OR USE THEM IN A SALAD, OR TO REPLACE THE OLIVE IN YOUR COCKTAIL.

TRY SOMETHING NEW

If you like cucamelons, why not give cantaloupe melons a try? They need an extremely warm and sheltered spot to grow, and are fairly high-maintenance, but the delicious fragrance and flavour of a homegrown cantaloupe makes it worth the work. Try the following two varieties:

'Ogen' – sweet, deep-yellow medium-sized fruit

'Sweetheart' – smallish fruit with orange flesh and grey-green skin

French bean *Phaseolus vulgaris*

French beans are an excellent addition to your veg-growing repertoire: they're tasty, easy to grow, and very generous croppers. The climbing varieties need plenty of room and something to scramble up, but if you only have a small growing space, there are also plenty of popular dwarf varieties that can be grown in containers.

SOWING

Choose your French bean varieties according to how much space you have: bush varieties for pots, and climbers with support for beds.

Start under cover in late March or early April. Plant each large bean in its own small pot or cell filled with seed compost, and cover it to a depth of 1cm (½in) with more compost or vermiculite. Water and leave in a warm place – a propagator or a windowsill. Seeds should germinate within ten days.

GROWING AND CARE

Pot up when the seedlings get too large for their first pots, but keep under cover until early June. Then harden off for a week, either in a cold frame, or simply by taking the pots outside during the day and bringing them back under cover at night, before planting them outside.

French beans like fertile soil, and lots of sun and water. In containers, use a compost mix containing plenty of organic matter; in a bed, dig in plenty of compost.

Dwarf beans can grow without support; climbers will need a trellis or wigwam to scramble up, or you can make your own supports (see pages 44-45). Plant the seedlings at the base of the support and train them up, tying in as necessary at the start (they'll soon attach themselves).

AS WELL AS STORING THEM DRIED, YOU CAN FREEZE A GOOD HARVEST OF FRENCH BEANS.

HARVESTING

Pick regularly once the beans are around 10cm (4in) long; the more you pick, the more will grow. Dwarf beans tend to have a slightly shorter cropping season than climbing varieties. If you have a larger crop than you can cope with, you can save some to dry as haricot beans.

THREE TO TRY

'Golden Gate' – large climbing variety; produces a big crop of flat, creamy-yellow pods

'Kenyan Bean' – dwarf variety; early to flower and crop – deep green beans

'Purple Teepee' – compact dwarf variety; magenta flowers are followed by deep purple, stringless pods

Sugarsnap and mangetout peas *Pisum sativum*

Homegrown peas are an all-round favourite. If you haven't grown peas before, start with the smaller types – sugarsnaps and mangetout. The compact plants crop well, and work in containers if you have limited space. You can help yourself to the ultimate fresh snack directly from the plant while you're picking.

SOWING

Sugarsnap varieties have rounded pods, mangetout flatter ones, but both are sweet and crisp, and both can be sown under cover or directly outside into a container or bed.

Under cover, start in March. Sow the peas singly into small pots or cells filled with seed compost, cover them with 1cm (½in) of compost, water well, and leave in a warm place such as a propagator or windowsill. They should germinate in five to ten days. Move outside when the plants are 15cm (6in) tall – they'll be less susceptible to slugs, which love the younger shoots, after any chance of a late frost.

Outside, sow in May. If you're planting in a large container, fill it with soil enriched with some homemade compost or well-rotted manure; if into the ground, rake the soil until fine. Make a circular drill (planting in a circle makes it easier to support the plants later on), then drop the seed peas in, spacing them 10cm (4in) apart and covering with 5cm (2in) of soil. They will take up to two weeks to germinate. Thin out the seedlings if necessary.

CHECK YOUR PLANTS CAREFULLY:
PODS HIDDEN AMONG LEAVES
CAN BE HARD TO SPOT.

GROWING AND CARE

Sugarsnaps and mangetout need regular watering. They also need light support: a cane tripod (tie three canes at the top) or a twiggy structure made with peasticks will do the job. If the birds start to eat the pods, drape garden mesh or netting over the support structure.

HARVESTING

Snip the peas from their stems, harvesting on the young side for the best flavour; pods will grow larger but also tougher if they're left too long on the vine.

THREE TO TRY

'**Cascadia**' – sugarsnap pea that delivers a very heavy, early crop

'**Shiraz**' – mangetout variety with pretty, pink-and-magenta flowers, followed by deep purple pods

'**Sugar Ann**' – sugarsnap type, with an exceptionally sweet flavour

Radish *Raphanus sativus*

With their pink or red colouring and their lovely spicy crunch, radishes are one of the most welcome spring-into-summer crops. They're quick to germinate, and fast to grow – with staggered sowings, you can dine off a single packet of seeds until midsummer, or choose several alternative varieties to compare, contrast, and find your own favourite.

SOWING

Sow the seed directly where you want to harvest your radishes (moving small seedlings around can cause splits in the roots as they grow). They're a good crop for containers or even window boxes, as well as in the ground. They prefer a moderately sunny spot – too much sun can mean luscious leaf growth at the expense of the crop underground. Scatter seed very thinly into compost enriched with a few scoops of well-rotted manure if you're growing in a container, or in rows into raked drills around 25cm (10in) apart if you're planting direct into the ground. Cover over with around 1cm (½in) of soil, then water sparingly and keep moist (not wet). Germination usually takes between seven and ten days.

GROWING AND CARE

When the seedlings have developed their second leaves, thin them out, leaving plants around 5cm (2in) apart. Eat the trimmings as a microgreen treat; they taste like spicy cress. Radishes are best repeat-sown in small quantities – a mini-crop sown every ten days should keep you supplied across spring and through summer.

HARVESTING

Around three weeks after germinating, your radishes should be ready to harvest. Pull up just one first – if it's not quite ready, leave your crop for another few days before trying again. Harvest radishes while they're young, crisp and on the small side; larger, older ones can become woody. Try them French-style, with butter and salt, to bring out their flavour.

RADISHES GIVE QUICK RESULTS, SO ARE A GOOD CHOICE FOR CHILDREN TO GROW.

THREE TO TRY

'**French Breakfast 3**' – cylindrical pink radish with a long, white root

'**Marabelle**' – small-leaved, compact plant with classic round, red fruit

'**Sparkler**' – smallish, round, bright pink radishes with a white root

Sweetcorn *Zea mays*

The taste of sweetcorn that's been picked an hour before you eat it is unforgettable, and easily earns it a place on the best-homegrown list. There's one proviso, though – sweetcorn plants are big, and you need to grow a block of them, so you'll need a medium-sized bed to raise them.

SOWING

To make sure you have the space to grow sweetcorn, do the maths before you start. Sweetcorn plants are wind-pollinated: the pollen blows from the male flowers, at the top of the plant, onto the female ones below that, once pollinated, develop into cobs. This means you need to plant in a block rather than a row, so that pollen will blow within it, plant to plant. Calculate on a minimum block of nine plants (with the plants producing an average of two cobs each, that's a projected 18-cob harvest), on a grid allowing 60cm (2ft) between each – around 2.5m (8ft) square in all.

Space planning done, start your sweetcorn under cover in April, sowing the seed kernels singly. Use individual pots of 12.5–15cm (5–6in) diameter filled with seed compost. The pots are slightly larger than usual, because the small seedlings hate their roots being disturbed. Cover over with 1cm (⅓in) compost, water, and leave in a warm place such as a propagator or windowsill. Seeds should germinate within ten days. Keep under cover until the nights are warmer and there is no risk of frost.

THE GRASS-LIKE, EARLY SHOOTS OF SWEETCORN QUICKLY DEVELOP INTO LARGE, STURDY PLANTS.

GROWING AND CARE

In June, plant young plants out in a block, as described above, into a raked bed to which plenty of organic matter has been added. Mulch around the base of the plants with well-rotted compost, and water regularly in dry weather.

HARVESTING

When the tassel at the top turns brown, the corn is ripe. Harvest cobs by twisting them off their stems shortly before you want to cook them.

THREE TO TRY

'Earlibird' F1 – super-sweet variety; cobs mature quickly for an early crop

'Lark' – crops mid-season; will tolerate slightly lower temperatures than some other varieties

'Swift' F1 – slightly shorter than other varieties; good reliable cropper

CHAPTER FOUR

SAVING & SWAPPING

Growing from seed can be addictive, and there are plenty of ways you can extend your new enthusiasm. If you've had a great year in the garden, you can complete the cycle of grow-your-own by collecting seeds from your plants to sow next year – it's easy when you know the basics. Gathering and labelling your harvest, and creating the beginnings of your own seed library are fun to do, too, but you may find that you've collected many more seeds than you think you'll need for your own space – so the next step is to share.

Visit a seed swap event and meet other gardeners to exchange with, or simply swap informally with friends. Alternatively, turn your seeds into gifts by making seed bombs with the seeds of some favourite flowers, and seek out opportunities for a bit of – strictly legal – guerrilla gardening around your neighbourhood. Read on to find out how.

SAVING SEED FOR NEXT YEAR

There are two main methods for saving seeds, depending on what kind of plant they come from, but neither is complicated or difficult. The important points to remember are that the seeds have to be fully mature before they are ready for you to collect them, and that it's essential to dry them really thoroughly before storing them, or they will go mouldy and spoil.

SAVING DRY SEED

The seeds of most annuals are 'dry' – that is – they mature on the plant inside seedheads and drop onto the ground when they're ready. Larger seeds can be picked up from the ground – nasturtium (*Tropaeolum majus*) seeds, for example, look like wrinkled little peas and are immediately recognizable. The seedheads of hollyhocks (*Alcea rosea*) have the seeds arranged in a circle inside; when they start to open, you can empty the dry seeds out. Keep some brown paper bags to hand to collect some of the smaller seeds. You can cut whole, dried seedheads off plants such as cosmos (*Cosmos bipinnatus*) or marigolds (*Calendula*), leaving a short stem, then put them head-first into a bag and shake the seed out without losing any. Check the seeds are ready first by crushing a seedpod between your fingers: it should feel dry and brittle. Empty the bag onto a plate and gently separate the seeds from the seedhead debris, the chaff.

DRIED MARIGOLD
FLOWER HEADS

MARIGOLD SEEDS,
SHAKEN FROM THE
DRIED FLOWER HEADS

READY TO GROW
NEW PLANTS THE
FOLLOWING YEAR

HYBRID SEEDS

Remember that it's not worth saving the seeds of F1 or F2 hybrids – many are sterile, and any that do germinate won't match the qualities of their parent plants.

BUTTERNUT SQUASH SEEDS

Flowers with complex, many-petalled mop-heads, like this hybrid marigold, produce very little seed to harvest.

HYBRID MARIGOLD

SAVING WET SEED

'Wet' seed is seed that's stored in pulp, in crops such as tomato and squash. Scrape the pulp out of a very ripe fruit (a teaspoon is useful when you're taking it out of a ripe tomato), put it in a sieve and rinse it, then drain the seeds on kitchen towel before leaving them to dry thoroughly on a plate (if you leave them on the towel, they'll stick to it as they dry).

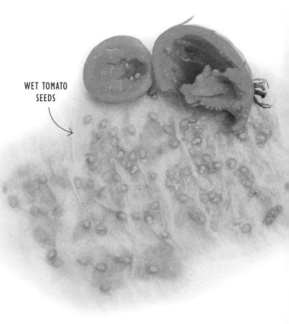

WET TOMATO SEEDS

Alternatively, for tomatoes, you can use the fermentation method – enthusiasts believe that this gets rid of possible moulds and pathogens, and improves the germination rate. Tip the pulp into a jam jar and add 5cm (2in) of water. Leave for two days (not longer, or the seed may start to rot), by which time there will be mould on top of the water, then sieve and rinse the seeds, and dry them as before.

START A SEED LIBRARY

When all your seeds have been gathered and thoroughly dried out, it's time to store them for next year. First, you need to put them into envelopes and make sure they're accurately labelled. Then, you need a cool, dry and dark place to keep them until it's time to start sowing again.

STORING SEEDS

Seeds should be completely dry before storing, so leave them out for an extra day or two if you're not certain. They should be somewhere airy, but not near a radiator or any other high heat source – they need to dry at room temperature. When they're ready, it's time to bag them up and label them. Tiny envelopes in either brown paper or glassine are easy to find online (glassine is transparent, coated paper, allowing you to see the seeds inside); either works well for seed storage.

If you have a lot of the same sort of seeds, you can use them for seed swaps, so bag them in several separate envelopes.

Label the envelopes clearly with the plant name, the cultivar, and the date the seeds were collected. If you're planning on attending seed swaps (see pages 134–135), then you may want to divide your seeds between a higher number of envelopes at this stage, rather than putting a lot together in one envelope.

STARTING A SEED BOX

To store your starter seed library, you need a sturdy tin or cardboard box (a medium rectangular box, shoebox-sized or slightly smaller, works well), which can be dedicated to your seed collection. You can buy attractive, purpose-made tins and boxes in nurseries and garden shops, or online, but some gardeners prefer to get creative and decorate their own with lettering, collage, stickers and so on (equally, of course, if you want to leave it plain, that's up to you). Whichever you choose, do label it clearly.

If you already have a big stack of different envelopes of seed to sift through, make card box dividers to create sections – start with 'Flowers', 'Vegetables' and 'Herbs', and then, as the collection grows, you can add some subdividers if needed. Make sure that your

seed store stays dry, either by putting a few spoonfuls of uncooked rice in an envelope and keeping it in with the seeds (it will absorb any moisture), or by saving and using the small sachets of silica gel that come with many mail orders, which will do the same job.

Make sure your seed box is stored somewhere cool and dry – in a shed or mini-greenhouse outdoors, or indoors, in an unheated room.

SEED SWAPS & PLANT FAIRS

One of the great pleasures of gardening is chatting with other enthusiasts, whether you're sharing the thrill of a great tomato harvest or bemoaning the aphids on your nasturtiums. Seed swaps – events when seed fanciers meet up to exchange seeds from their most successful plants – have become incredibly popular in recent years; it's well-worth seeing if there are any held locally.

WHAT TO EXPECT

Seed-swap events vary a lot in size, from a stall in a corner of your local summer fair to large, dedicated gatherings, such as the annual 'Seedy Sundays', which host dozens of swappers – and sellers – including some commercial stands. The larger seed swaps include every kind of gardener, from the back-garden grower or allotmenteer, keen to exchange a few packets of their most successful sunflowers, squashes or courgettes, to much larger setups and groups offering a very wide choice of seeds.

JOIN THE CLUB

If you want to take part in a seed swap, find a small event near you, and approach the organizers about taking a stall. If you only have a few seeds to offer, you may be able to join a community stall with other small-scale growers. Make sure your packets are marked with the variety and harvesting date and, ideally, add any useful growing information you can share.

ORGANIC AND HERITAGE SEEDS

Organic seeds are exactly as they sound: seeds from plants that have been sustainably grown without the use of any pesticides or chemical fertilizer. You'll see them both in catalogues and at seed fairs. Many gardeners believe that they are both ethically and practically the best option for home growers.

Heritage seeds, sometimes also called 'heirloom' seeds, are a less straightforward subject. The label refers to old varieties, mainly vegetables, most of which exist outside the national list of approved seed types – that is, seeds that are licensed for commercial sale. Some heritage varieties have made it through onto the national list and can be bought from seed companies, but you won't find the majority for sale in a catalogue. Enthusiasts champion them, claiming that they're more flavourful and suitable for small-scale growing. Although most can't be sold, they do often pop up at seed swaps, offered as swaps so that no money changes hands. The stories of a lot of heritage varieties are fascinating, and those who grow them are often true seed experts, so if you spot seeds with a heritage label, it's definitely worth chatting to the stallholder.

GREEN YOUR SURROUNDINGS

The more time you spend sowing and growing outside, whatever space you have to garden in, the more you tend to become aware of how good it is for you. Look around for wider opportunities to pass the enthusiasm on. Seed bombs are a great present for gardening or even non-gardening friends. Stay aware of your surroundings, too: it's easy to spot places where a few seeds might make a difference.

Over the last few years, it's become increasingly frequent for gardeners living in towns and cities to 'green' some corners of their environment – it's their efforts you're noticing when you spot a tree pit surrounded by flowers rather than the usual bare earth, for example. Follow their lead and use some of your surplus flower seeds to scatter – either loose or in the form of small seed bombs – where they stand a chance of growing, and adding colour to an otherwise dreary spot. Only use seed from garden flowers you've grown yourself, and don't introduce them anywhere inappropriate – this should be more stealth-sowing than guerrilla-gardening. The reward is to see, the following season, a handful of flowers blooming in a previously forgotten corner.

BAG UP SEED BOMBS FOR GIFTS ALONG WITH SOME FRESH FLOWERS.

HOMEGROWN SEED BOMBS

Seed bombs are handcrafted balls of hardened soil with seeds inside. Throw them into a neglected corner and, in time, they'll produce a surprise crop of flowers. Make them with flower seeds that you gathered yourself from plants that you grew, and they make a uniquely personal present. They're easy enough to make with children if you want a craft project that ties in with growing activities they've already had a part in.

To make seed bombs, you need some of the same seed compost that you use for sowing seeds, and a packet of clay powder (you can buy this in any craft shop), plus a palmful of flower seeds you've grown. Mix up a bowlful of two parts compost to one part clay powder, and pour in a little water. Start with very little water and add as necessary; you're aiming for a mixture that's sticky, but that will hold together when you shape it. Scatter the seeds in and mix everything thoroughly with your hands. Then, shape the mixture into balls. Traditionally, seed bombs are around the size of a tennis ball, but you can make miniature ones, more golf-ball sized, too. Leave, spaced, on sheets of newspaper – they will take a few days to dry out thoroughly.

Only use your own flower seeds in seedbombs – that way, there's no chance that you'll introduce unwanted or invasive species.

GLOSSARY

ANNUAL
A plant that completes its life cycle in one year, from germination to maturing, flowering, setting seed and dying.

BIENNIAL
A plant that completes its life cycle in two years, germinating and growing foliage in the first year, and flowering, setting seed and dying in the second.

BOLT
A tendency in a plant to bolt means that it runs to seed early, cutting short its flowering and/or cropping season. Sometimes plants with this tendency will do better when grown out of full sun and in moister soil.

CLOCHE
A clear dome or frame placed over a plant to protect it when it's planted in the earth. Can be made from glass, plastic, or as a metal frame with glass panels.

COLD FRAME
A deep frame, usually made from wood, with a hinged, clear lid of plastic or glass, but without a base, used to protect plants that need cover outdoors.

CORDON
A method of growing, which trains a plant along a single stem, taking off side growth as the plant grows, so that it fruits up that stem. In vegetable growing, it's usually used to refer to a type of tomato and the way that it's grown, but it can refer to other vegetables and fruits, too.

CULTIVAR
The condensed form of 'cultivated variety', which describes a plant that has been bred selectively to enhance desirable qualities.

DEADHEAD
To cut off the spent flower heads on a plant, prompting it to produce more flowers.

DORMANT
Used to describe a seed that's not in an active state of growth or germination.

DRILL
A marked-out furrow in a prepared bed, ready for sowing seed.

GERMINATION
The sprouting of a seed that was previously dormant.

GLUT
A harvest of one kind of crop that ripens all at once, giving the grower far more than they need in one go!

HALF-HARDY ANNUAL
An annual that can be grown outside but that is too tender to survive very cold weather.

HARDENING OFF

The process of toughening up a plant gradually when moving it outside, usually by leaving it in a cold frame or unheated conservatory for a few days, or taking it outside for a few days but bringing it back in overnight.

HARDY

A plant that is capable of tolerating low temperatures, including frost.

HEIRLOOM SEEDS

Seeds from plant varieties that are at least 50 years old. The term 'heritage' seeds is used interchangeably with 'heirloom'.

HYBRID

A deliberately engineered cross between two cultivars of a plant, made to produce offspring that have the more desirable characteristics of both parent plants.

ORGANIC SEEDS

Seeds that have been produced by a grower who runs their business along certified organic lines. Parent plants will not have been treated with synthetic chemicals, and seeds will not have been treated with any chemicals after harvest.

PERENNIAL

A plant with a life-cycle of three years or more. It may die back, shed its leaves, or go into dormancy in the colder months.

POT ON

Transferring growing seedlings from the small containers in which they germinated into bigger ones.

SEED COMPOST

Compost sold especially for sowing seeds. It's sterile, so new seeds won't be affected by any of the micro-organisms and fungi in regular potting compost, but it doesn't contain nourishment for plants after seeds have germinated and are actively growing.

STRATIFICATION

The process that some 'fussier' seeds need in order to germinate, which imitates the natural conditions in which the parent plants grow. It usually calls for periods when the seeds are kept at specific, usually low, temperatures.

THIN OUT

The process of weeding out and discarding the weaker seedlings in a group, leaving the stronger ones to grow.

VERMICULITE

A naturally light mineral, sold in bags, which is used to aerate soils that are too heavy for some seeds, and to cover seeds before germination.

FURTHER RESOURCES

The RHS website should be your first port of call – it offers a mass of free information on growing from seeds, whatever your gardening circumstances. Check it out at The Royal Horticultural Society (RHS) www.rhs.org.uk

The RHS also offers a seed scheme to members, giving them the exclusive chance to buy seeds harvested from RHS gardens.

BOOKS
The Flower Garden: How to Grow Flowers from Seed by Clare Foster & Sabina Rüber (Laurence King Publishing, 2019)

Growing from Seed (RHS Practical series), (Dorling Kindersley, 2002)

RHS Grow for Flavour by James Wong (Mitchell Beazley, 2015)

RHS Plants from Pips by Holly Farrell (Mitchell Beazley, 2015)

RHS Propagating Plants by Alan Toogood (Dorling Kindersley, 2019)

RHS Seeds: The Ultimate Guide to Growing Successfully from Seed by Jekka McVicar (Kyle Cathie, 2012)

WEBSITES
As well as selling seeds, many seed companies offer extensive online resources, too. Some of the names below will be familiar from the packets you see in sales racks; others, including a lot of the smaller companies, sell exclusively online. Every company has its own personality – enjoy hunting out your own favourites.

Chiltern Seeds
chilternseeds.co.uk
Wide range of flower, vegetable and herb seeds, including some novelties and unusual varieties. Produces a full printed catalogue with lively text and lots of extra features.

Emorsgate Seeds
wildseed.co.uk
Specialists in wildflower seeds, with a wide range of very unusual varieties, some not found elsewhere – perfect if you want to sow your own mini-meadow or wildflower patch.

Higgledy Garden
higgledygarden.com
Characterful one-man business selling flower seeds, with a concentration on wildflowers and flowers for cutting gardens. Online growing guides, and irresistible seed packets handprinted with a 1970s-style flower-power design.

Organic Gardening Company

organiccatalogue.com
Exclusively organic range, listed online and as a paper catalogue, including vegetable, flower and herb seeds.

Pennard Plants

pennardplants.com
Small Somerset-based nursery run from a Victorian walled garden. Offers a large range of heritage varieties, including flowers, herbs and vegetables, served up in quirky illustrated seed packets.

Plant World Seeds

plant-world-seeds.com
Easy-to-navigate site with a wide choice of seeds – flowers, vegetables and herbs – including hybrids and some unusual varieties.

Real Seeds

realseeds.co.uk
Small company specializing in old-fashioned vegetable varieties and a few flowers (no hybrids). Engaging descriptions accompanied by lots of very useful growing information.

Robinson Seeds

mammothonion.co.uk
Long-established company offering a good range of high-quality vegetable seeds.

Seedaholic

seedaholic.com
Irish-based company with an extensive catalogue of flowers, vegetables and herbs, including a large selection of organic seeds. Packets are sent with detailed individual information sheets.

Suttons

suttons.co.uk
Large company whose seeds are sold in most nursery outlets. Offers a comprehensive range of flower, vegetable and herb seeds, with full catalogues that can be downloaded – or ordered by post if you prefer a paper copy.

Thomas Etty Esq.

thomasetty.co.uk
Eccentric site written entirely in the style of a Victorian plantsman. Extensive online catalogue of flower and vegetable seeds (including many old-fashioned varieties) which is an enjoyable as well as a useful read.

Thompson & Morgan

thompson-morgan.com
One of the giants, with a huge choice of flowers, vegetable and herbs. Comprehensive catalogues that can be digitally downloaded or placed on order if you prefer a paper copy.

INDEX

IMAGE CREDITS

FLOWER SOWING GUIDE

	JAN	FEB	MAR	APRIL
Hollyhock biennial*/perennial		●	●	
Bishop's flower				●
Snapdragon	●	●	●	
Columbine	●	●		
Cornflower			●	●
Honeywort		●	●	
Cup-and-saucer vine		●	●	
Coreopsis			●	●
Cosmos			●	●
Foxglove biennial*/perennial			●	●
California poppy			●	●
Sunflower			●	●
Sweet pea		●	●	●
Ox-eye daisy			●	●
'Poached egg' plant		●	●	●
Alyssum			●	●
Tobacco plant		●	●	●
Love-in-a-mist			●	●
Nasturtium			●	●
Zinnia			●	●

Biennial*: sow one year to flower the next